LEST INNOCENT BLOOD BE SHED

LEST INNOCENT BLOOD BE SHED

THE STORY OF THE VILLAGE OF LE CHAMBON AND HOW GOODNESS HAPPENED THERE

Philip P. Hallie

HarperPerennial
A Division of HarperCollins*Publishers*

To
Elaine Norych Geller,
who, after it all, prevails

Lines from "In Memory of W. B. Yeats" reprinted from *Collected Poems* by W. H. Auden, edited by Edward Mendelson, by permission of Random House, Inc.

A hardcover edition of this book was published in 1979 by Harper & Row, Publishers, Inc.

HarperCollins books may be purchased for educational, business, or sales promotional use. For information please write: Special Markets Department, HarperCollins Publishers, Inc., 10 East 53rd Street, New York, NY 10022.

First Harper Torchbooks edition published 1985. First HarperPerennial edition published 1994.

Library of Congress Cataloging-in-Publication Data

Hallie, Philip Paul.

Lest innocent blood be shed : the story of the village of Le Chambon and how goodness happened there / Philip P. Hallie. — 1st HarperPerennial ed.

p. cm.

Originally published: New York : Harper & Row, c1979. With new introd.

Includes bibliographical references and index.

ISBN 0-06-092517-5 (pbk.)

1. Jews—France—Le Chambon-sur-Lignon. 2. World War, 1939–1945—Jews—Rescue—France—Le Chambon-sur-Lignon. 3. Holocaust, Jewish (1939–1945)—France. 4. Le Chambon-sur-Lignon (France)—Ethnic relations. I. Title.

DS135.F85L434 1994

944'.813—dc20 93-45254

---

05 04 03 RRD 30 29 28 27 26 25

To
Elaine Norych Geller,
who, after it all, prevails

Contents

PART FOUR: CONSEQUENCES—1943–1944

PART FIVE: THE ETHICS OF LIFE AND DEATH

Photographs follow page 116.

ACKNOWLEDGMENTS

This book is full of the names of people who helped make it. Still, some people who made an immense contribution, too great for a brief description, do not appear in the text; and so I shall simply list their names, leaving it to them and me to know how much I owe to them: John J. Compton, Erwin A. Glikes, Barbara Grossman, Daniel Stern, and Gloria Stern.

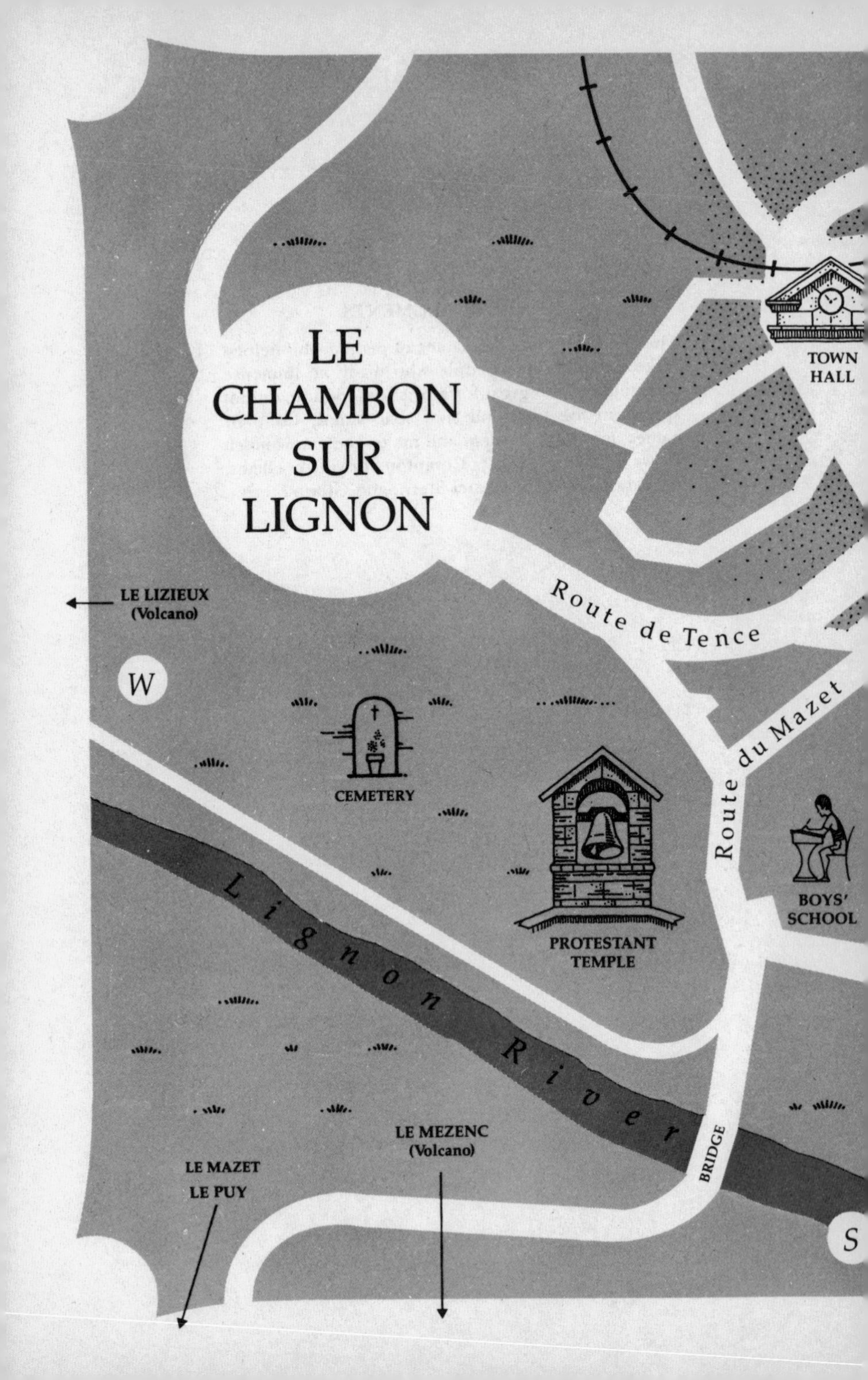

LE CHAMBON SUR LIGNON
TOWN HALL
Route de Tence
LE LIZIEUX (Volcano)
W
CEMETERY
Route du Mazet
Route
PROTESTANT TEMPLE
BOYS' SCHOOL
Lignon River
BRIDGE
LE MEZENC (Volcano)
LE MAZET
LE PUY
S

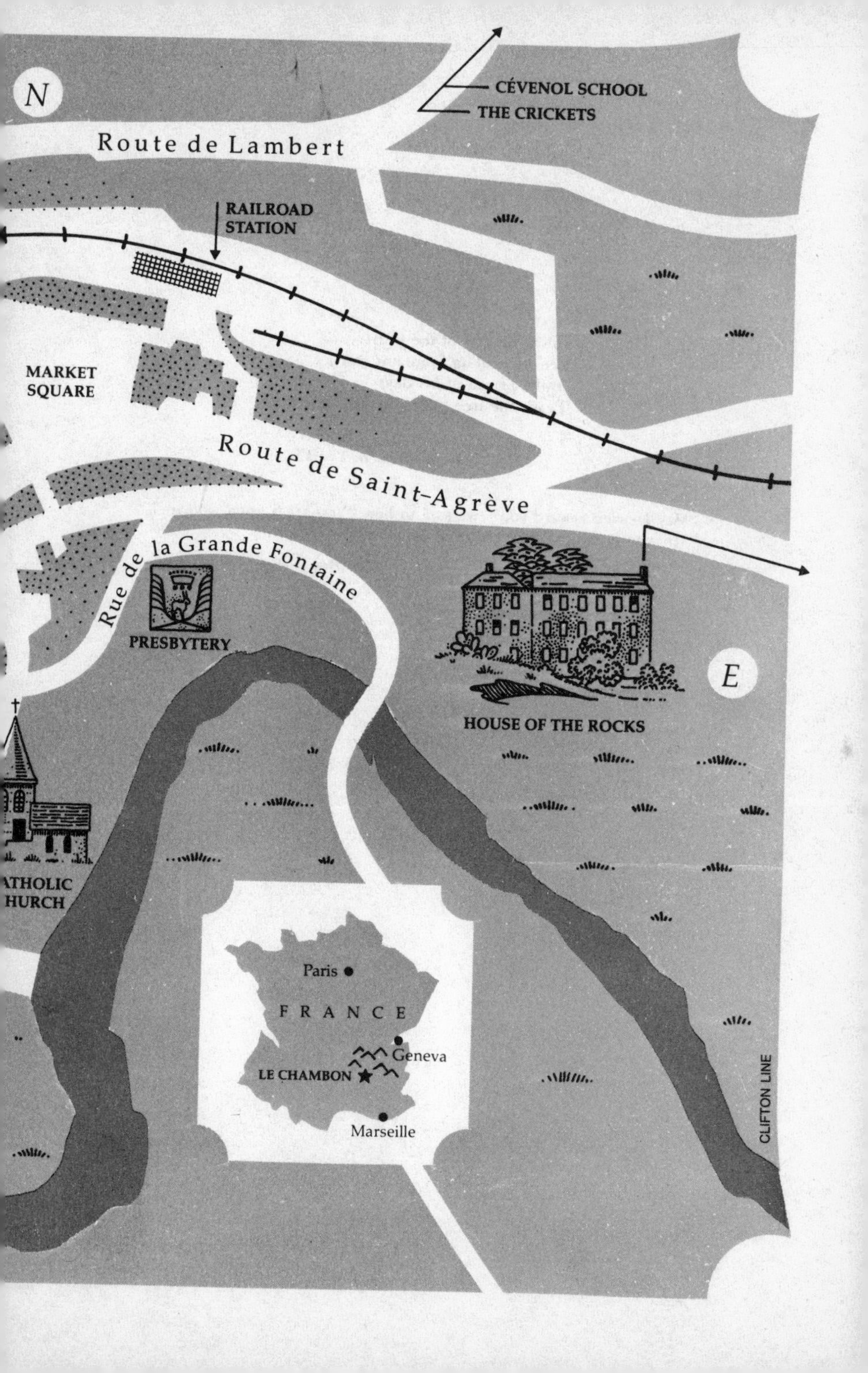
N
CÉVENOL SCHOOL
THE CRICKETS
Route de Lambert
RAILROAD
STATION
MARKET
SQUARE
Route de Saint-Agrève
Rue de la Grande Fontaine
PRESBYTERY
HOUSE OF THE ROCKS
E
ATHOLIC
HURCH
Paris
FRANCE
Geneva
LE CHAMBON
Marseille
CLIFTON LINE

In the deserts of the heart
Let the healing fountains start,
In the prison of his days
Teach the free man how to praise.

—W. H. Auden,
"In Memory of W. B. Yeats"

"May Heaven reward you!" we said to her. "You are a good woman."
"Me, young ladies?" she returned with surprise. "Hush!"

—Charles Dickens, *Bleak House*

Introduction to the HarperPerennial Edition

1.

This is a local story about a sequestered French mountain village. Its name appears in very few history books, mainly because the story concerns only a few compassionate people, a pittance compared with the millions and millions of people who were killing or dying upon the continent of Europe during World War II while the story of the village was taking place. During the four years of the German occupation of France, the village of Le Chambon, with a population of about three thousand impoverished people, saved the lives of about five thousand refugees (most of them children). The moral brilliance of the villagers does not light up the moral darkness around the village as much as it makes that vast darkness seem darker by contrast.

When I wrote the story of Le Chambon I did not believe this.

The story of the Huguenot village of Le Chambon all but filled my life, so that there was little room in me for the vast, murderous story of stories called "World War II." I had lived in the village for a few months in the seventies—long after the war—and I had spent many hours interviewing all of its surviving leaders. As a result I came to love Magda Trocmé, the Italian widow of the pastor of the village during the war years. With her almost raucous voice she was a confluence of little faults and immensely powerful virtues, like some of the unfinished sculptures of her countryman, Michelangelo. She was my main informant. I found myself dreaming about her and I wrote *Lest Innocent Blood Be Shed* as a sort of narrated eulogy to her as well as to the goodness of her fellow villagers. Whatever else the book was, for me it was personal, not a set of facts that should be told in objective abstractions, like statistics.

Shortly after the book came out, I received among my first "fan letters" a letter that reminded me painfully that I had indeed written a personal, a local, a little story. The letter went:

> Dear Professor Hallie,
>
> There is only one important thing to say about the Holocaust. It was merely a geological-type almost inanimate event (physical event). No one was responsible. No one started it. No one could stop it.
>
> Le Chambon wasn't even in the war. Nothing happened west of National Route 7 in southern France. The obscurity should be an insight. Reverend Trocmé [the leader of the village] has a miniscule number of equally eccentric kindred-spirits. . . .

And it went on to say that only vast forces like great armies "make history," make and break human institutions. The story of a few nonviolent eccentrics who did nothing to stop Hitler's armed forces mattered only to a few mushy-minded moralists like me.

At first, thinking that this was only the first of many similar reactions to my book, I started sketching out arguments against it. One of my arguments went: "Vast institutional cruelties exist because people like you believe that flesh-and-blood individuals can do nothing that counts. Well, *something happened* west of National Route 7 in southern France. Real people with their own proper names saved real human beings in that village. And these precious few people count."

But in the end I came to feel that there was an abyss between the writer of that letter and me, an abyss that no words could bridge. And so I wound up sending my first fan a picture post-card of Le Chambon with a few words on it, "Thanks for your point of view. Still, something really happened here."

I found myself wondering whether Harper & Row knew what they were doing when they published my book. After all, the history of World War II could be written in great detail and with great accuracy without mentioning Le Chambon. How could a wide public find such a little story important enough to read?

But then many more letters came, and none of them belittled the village. The one letter that stands out from all the rest came from a seventeen-year-old girl, whose parents had just given her *Lest Innocent Blood Be Shed* as a birthday gift. Referring to her parents she wrote:

> ...They saw that I was becoming depressed from my reading about the Holocaust...and they wanted to give me a gift of hope, of reaffirmation of what is best in people.
>
> ...I have finished the book now, and I wish I could tell you how much it has helped me. Every time I will look at your book in my bookcase, it will be like hearing ... two words.... Don't cry!
>
> I would give you my return address but then you would maybe feel obliged to answer me, and that would be time taken away from your teaching and your writing of books, which is where I feel you give yourself the most. So, I shall just give you my name, and my love.

Vast power that defends and destroys great institutions is real, and the heartfelt tears of a young woman are real. There is force and there is validity in both of these letters, as well as narrowness. I had been a combat artilleryman in the European theater and I knew that decent killers like me had done more to prevent the mass murders from continuing than this pacifist mountain village had done. And so I found myself wavering between praising military valor above all and praising moral valor above all. I could easily make these two points of view consistent with each other (one was a "public" perspective and the other was a "personal" perspective, etc.), but the questions that kept gnawing at me were: Where does your heart lie? Which of these two letters counts the most for you? Who, in short, or what, are you?

The answers to these questions did not come from searching my soul for my ultimate loyalty; nor did it come from neat, abstract philosophical distinctions between public and private points of view. It came from an image, a vision if you will, that was granted me during a lecture-discussion that occurred a few years after the publication of *Lest Innocent Blood Be Shed.*

One evening I was speaking to a group of women in the assembly room of a large hotel in Minneapolis. The women were all leading fund raisers for the United Jewish Appeal. They were a formidable audience, with their intense eyes and their energetic personalities. I was talking to them about the killing of more than a million children by the Nazis, and about the village of Le Chambon as the safest place for children on the continent of Europe during the war years.

When my lecture was over, I asked for questions or comments. A woman in the back of the room stood up and asked me if the village of Le Chambon was in the Department of Haute-Loire in south-central France where the great Loire River has its origin. Her accent marked her as French, and so did her knowledge of the fact that there is more than one village in France called Le Chambon. I told her that indeed the Le Chambon I was talking about is in the Haute-Loire.

She was a powerful woman wearing a sheath dress that made her body look like a slender cannon, taut, full of explosive power. But for a moment the cannon seemed to crumple. She stood there silent for what seemed to be a long time and then she said, "Well, you have been speaking about the village that saved the lives of all three of my children."

There was absolute silence. She drew herself erect in that sheath dress, and she said in formal tones, "I want to thank you for writing that book. Now that the story is in print my American friends who read your book will understand those days better than they have. You see, you Americans live on an island, and though your people fought and died in the war, you see the Second World War from a distance, from a distance. . . ."

There was another silence as she stood there. She had something more to say, but apparently she was trying to decide whether to say it. Then she asked if she might say one more thing. Nobody had the breath or the authority to say a word; she was in total command.

She came to the front of the room, turned to face the audience, and said, "The Holocaust was storm, lightning, thunder, wind, rain, yes. And Le Chambon was the rainbow."

A few people in the room gasped, while she and I looked at each other, and I said, "The rainbow," and she nodded slowly.

We understood each other. We understood that the rainbow is one of the richest images in the Bible. The rainbow is the sign God put up in heaven after the great Flood. The sign meant: ". . . never again shall all flesh be cut off." In His explanation of the rainbow, God repeated the phrase "never again," and ever since the Holocaust, Jews have been repeating that phrase.

The rainbow reminds God and man that life is precious to God, that God offers not only sentimental hope, but a promise that living will have the last word, not killing. The rainbow means realistic hope. For that woman whose three daughters were saved by the villagers of Le Chambon, history is not hopeless, because of the unshakable fact that lives were saved in Le Chambon.

Ever since the woman from Minneapolis witnessed to that hope, I realized that for me too the little story of Le Chambon is grander and more beautiful than the bloody war that stopped Hitler. I do not regret fighting in that war—Hitler had to be stopped, and he had to be stopped by killing many people. The war was necessary. But my memories of it give me only a sullied joy because in the course of the three major battles I participated in, I saw the detached arms and legs and heads of young men lying on blood-stained snow.

The story of Le Chambon gives me an unsullied joy. Why?

2.

There are such things as plain facts that I will allow nobody to explain away or bully me into doubting. One of them is the story of Le Chambon as the testimony of hundreds of people who tell and confirm that story. But there is another type of fact that is as plain as deeds. It is the belief that it is better to help than to hurt. There are sociopaths and others who believe otherwise—though they themselves believe it is better for others to help *them* than to hurt *them.* But there are also people like me who believe that help is better than harm, that caring-for is better than hurting, and that this is the case most poignantly when we are thinking about murdering children or saving their lives. People like us may violate or compromise that belief in our actions and feelings, but we will never let anybody argue us out of believing it. It is not a matter of opinion for us; it is a fact, and a fact is something that is *known* to be true, known by somebody, known perhaps by many, but perhaps not known by everybody on earth. Such a fact is not as plain as a pikestaff, but no philosophical abstractions can undermine it for those who know it. It is more solid than opinions: it is rock bottom.

This moral fact was of little importance to my first "fan"; or perhaps it was not even real enough for him to be counted as a fact. But it is at the living center of every response I have received since that first letter, and it is at the living center of the story of Le Chambon during the four dangerous years of the German occupation of France. In the writing of that story I did not see the fact clearly; it took the responses to the book to make me see how important it was to me, to the story, and to all those who found the story *real.*

And the reason the rainbow is closer to my heart than the Flood that was World War II is that the people of Le Chambon helped without harming, saved lives without torturing and destroying other lives. This is why the rainbow gives me unsullied joy and necessary and useful killing does not.

I believed this while I was writing the story of Le Chambon, but I never saw it as clearly as I saw it after the book was published. The young woman's "Don't cry!" and the older woman's "And Le Chambon was the rainbow" helped me to see more clearly than I ever had what lies closest to my heart.

3.

To understand the story of Le Chambon is not merely a matter of understanding historical and moral facts. A fact is plain, even obvious, once you simply face it. It is part of the quotidian world, and when you know it, you simply know it; you are not mystified by it. But the story of Le Chambon has more than everyday factuality in it. It has something supernatural in it.

Almost from the beginning of my interviews with Magda Trocmé, the story of all those rescues had a feeling of mystery about it. In my early notes I keep finding the question: How could the village have survived for four whole years of the German occupation? The Gestapo knew about it—there was a limited Gestapo raid in the summer of 1943 and there was at least one informer who regularly sent fat packets to the puppet French government of the Germans in Vichy; packets that must have given many details about the rescue operation. For years the village was known by Germans and Frenchmen alike as "That nest of Jews in Huguenot country." In the afternoons a little steam train brought hundreds of Jews and other refugees into the village in broad daylight. And this went on for years.

In fact, the rescue operation was an open secret in the region and elsewhere in France. Why wasn't the village, or at least the rescue effort, crushed in the course of those four long years, when that effort involved saving the lives of the people most hated by the German government? Oradour-sur-Glane was destroyed and other villages in France suffered greatly for opposing the wishes—or even for seeming to oppose the wishes—of the German occupants. But aside from that one raid on the House of the Rocks and The Crickets, Le Chambon remained untouched by the most bitter enemies of the Jews, the Gestapo.

One possible answer is that the villagers were nonviolent. They did not pose a threat to the German military presence in the region. The commander of the German occupying troops in the Haute-Loire had as his main task keeping the peace, so that the Germans could milk France dry with a minimum of troops committed to the region. A quiet, pacifist village helped him perform that task. But the Gestapo was independent of the occupying troops and could override the wishes of the army with ease. It had a different task from that of the army: that of destroying all political and racial enemies of the Third Empire, especially the Jews. Members of the Gestapo knew that the village was full of Jews, knew it in detail, but they did not bother to round up all of the Jews in the village, though they were seizing and killing Jews in France during the whole four years of the Occupation! How could this be?

Long after *Lest Innocent Blood Be Shed* was published I was still asking myself this question. The facts I could gather before and after the book was published did not help me to answer it. And so one day I posed the question to a friend on the faculty of Wesleyan University. He was a distinguished mathematician and always a cool-headed, circumspect thinker.

The answer he gave stunned me: "It was a miracle."

Of course this was his way of saying that the survival of the village was beyond his comprehension, but miracles also have to do with God's power, or at least with man's belief in that power. His answer reminded me of the fact that the leader of the rescue efforts of the village was the Protestant Pastor André Trocmé, who was a passionately religious person. I could not say with confidence that I believed God had worked that miracle through Trocmé, but I had to say in all certainty that Trocmé's belief in God was at the living center of the rescue efforts of the village. My Wesleyan friend's surprising answer reminded me of the spiritual power of Trocmé and his parish, and of the importance of that power to the success of those rescue efforts over four long years. Somehow their faith was "a mighty fortress," as the old Protestant hymn goes.

And so I have come to believe that if a miracle is a marvelous event involving spiritual power at its vital center, the efficacy and the survival of the village were miraculous. Whether God used Trocmé and his fellow villagers as instruments, and whether God protected the village from destruction, I leave up to the theologians. But a belief in God certainly motivated Trocmé and the villagers, and the love the villagers displayed was indeed effective. For me and for others it was beautiful and beyond all quotidian understanding, like the rainbow after the Flood.

Philip Hallie
August 1993

Prelude

There was once an art critic, I have been told, who had a sure way of identifying ancient Maltese art objects: he found himself crying before them. John Keats had a similar reaction to excellence: the thought of his beloved Fanny Brawne, or of anything he associated with her, "goes through me like a spear," he said.

Of course, these are symptoms of an awareness of excellence. They are mere reactions, not rules that we ordinary people can use to separate excellent things from dross. But any doctor will tell you that symptoms are important, and just as pain can be a symptom of disease, painful joy can be a reliable reaction to excellence.

One afternoon I was reading some documents relating to Adolf Hitler's twelve-year empire. It was not the politics of these years that was at the center of my concern; it was the cruelty perpetrated in the death camps of Central Europe. For years I had been studying cruelty, the slow crushing and grinding of a human being by other human beings. I had studied the tortures white men inflicted upon native Indians and then upon blacks in the Americas, and now I was reading mainly about the torture experiments the Nazis conducted upon the bodies of small children in those death camps.

Across all these studies, the pattern of the strong crushing the weak kept repeating itself and repeating itself, so that when I was not bitterly angry, I was bored at the repetition of the patterns of persecution. When I was not desiring to be cruel with the cruel, I was a monster—like, perhaps, many others around me—who could look upon torture and death without a shudder, and who therefore looked upon life without a belief in its preciousness. My study of evil incarnate had become a prison whose bars were my bitterness toward the violent, and whose walls were my horrified indifference to slow murder. Between the bars and the walls I revolved like a madman. Reading about the damned I was damned myself, as damned as the murderers, and as damned as their victims. Somehow over the years I had dug myself into Hell, and I had forgotten redemption, had forgotten the possibility of escape.

On this particular day, I was reading in an anthology of documents from the Holocaust,[1] and I came across a short article about a little village in the mountains of southern France. As usual, I was reading the pages with an effort at objectivity; I was trying to sort out the forms and elements of cruelty and of resistance to it in much the same way a veterinarian might sort out ill from healthy cattle. After all, I was doing this work not to torture myself but to understand the indignity and the dignity of man.

About halfway down the third page of the account of this village, I was annoyed by a strange sensation on my cheeks. The story was so simple and so factual that I had found it easy to concentrate upon *it*, not upon my own feelings. And so, still following the story, and thinking about how neatly some of it fit into the old patterns of persecution, I reached up to my cheek to wipe away a bit of dust, and I felt tears upon my fingertips. Not one or two drops; my whole cheek was wet.

"Oh," my sentinel mind told me, "you are losing your grasp on things again. Instead of learning about cruelty, you are becoming one more of its victim. You are doing it again." I was disgusted with myself for daring to intrude.

And so I closed the book and left my college office. When I came home, my operatic Italian wife and my turbulent children, as they have never failed to do, distracted me noisily. I hardly felt the spear that had gone through me. But that night when I lay on my back in bed with my eyes closed, I saw more clearly than ever the images that had made me weep. I saw the two clumsy khaki-colored buses of the Vichy French police pull into the village square. I saw the police captain facing the pastor of the village and warning him that if he did not give up the names of the Jews they had been sheltering in the village, he and his fellow pastor, as well as the families who had been caring for the Jews, would be arrested. I saw the pastor refuse to give up these people who had been strangers in his village, even at the risk of his own destruction.

Then I saw the only Jew the police could find, sitting in an otherwise empty bus. I saw a thirteen-year-old boy, the son of the pastor, pass a piece of his precious chocolate through the window to the prisoner, while twenty gendarmes who were guarding the lone prisoner watched. And then I saw the villagers passing their little gifts through the window until there were gifts all around him—most of them food in those hungry days during the German occupation of France.

Lying there in bed, I began to weep again. I thought, Why run away from what is excellent simply because it goes through you like a spear? Lying there, I knew that always a certain region of my mind contained an awareness of men and women in bloody white coats breaking and rebreaking the bones of six- or seven- or eight-year-old Jewish children in order, the Nazis said, to study the processes of natural healing in young bodies. All of this I knew. But why not know joy? Why not leave root room for comfort? Why add myself to the millions of victims? Why must life be for me that vision of those children lying there with their children's eyes looking up at the adults who were breaking a leg for the second time, a rib cage for the third time? Something had happened, had happened for years in that mountain village. Why should I be afraid of it?

To the dismay of my wife, I left the bed unable to say a word, dressed, crossed the dark campus on a starless night, and read again those few pages on the village of Le Chambon-sur-Lignon. And to my surprise, again the spear, again the tears, again the frantic, painful pleasure that spills into the mind when a deep, deep need is being satisfied, or when a deep wound is starting to heal.

That night, I decided to try to understand all this. I decided to understand it so that I could hold it more firmly than one can hold a tear, or an image. Since I was a student and a teacher of ethics, I would use what I had learned about man's standards of ethical excellence to help me understand the blessing—at least for me—of Le Chambon. Those involuntary tears had been an expression of moral praise, praise pressed out of my whole personality like the juice of a grape. And part of that personality had been the ideas of goodness and of evil that I had been learning and teaching for decades.

But I was not going to make Le Chambon an "example" of goodness or moral nobility; I was not going to use this story to explain some abstract idea of ethics. Ends are more valuable than means; understanding that story was my end, my goal, and I was going to use the words of philosophical ethics only as a means for achieving this goal. Or, to be more, accurate, I was going to use the words of ethics to help me understand my deeply felt ethical praise for the deeds of the people of Le Chambon.

A year later, Pastor Édouard Theis was holding my arm to keep from slipping on an icy road in Le Chambon. The winter wind of the plateau, *la burle,* was blowing in its strange way: instead of pushing the snow away from us, off the plateau on which Le Chambon stands, it whirled the snow low and close around our feet. It hardly moved the tall pines on both sides of the road. It caught in its whorls the low, long-fingered evergreen bushes on the sides of the road. The bushes thrashed their green fingers around us, pointing in a thousand directions and in no direction at all: The people of the plateau used the long twigs to make their brooms in the old days, and so instead of calling the bushes by their proper name, *les genêts,* they called them *les balais* (brooms).

Night was almost at its darkest. The Protestant pastor and I were coming back from a day of interviews with some of the people of Le Chambon. Theis was miserable with influenza and still heavy with the sadness, two years old, of having lost his wife, Mildred. Everything—the wind, the slippery roads, the wild fingers of the broom—everything expressed loss, pointless loss, a whirling in deepening darkness. "Oh," Pastor Theis had said during one of our interviews, "I have been *unstable* since the death of Mildred Theis." And his heavy head, his big, curved nose that descended to a point, his wide shoulders—all seemed to be collapsing. This was the man who had helped another pastor, André Trocmé, make the village of Le Chambon one of the main forces in France for saving the lives of Jewish refugees during the four years of the Nazi occupation of France. In the last months of those years, when he was on the Gestapo's black list, he had guided refugees through the mountains of eastern France, through the French police and the German troops to the border of Switzerland and safety.

Though he reached for me to keep from falling on those icy mountain roads, he was helping me to keep from slipping as we walked together slowly. I had seen pictures of him in his thirties: the "rock of Le Chambon," the massive presence with a full, almost round face and big, heavy shoulders. But now Theis was seventy-five years old, and I felt under his coat (actually two thin, ragged raincoats) a thin, trembling man.

My God, I thought as we walked through *la burle,* we're losing everything. It's as if the broom were sweeping everything away, or inward and downward into darkness. At our last interview we had been talking with Madame Marion and her daughter, two of the strong women of Le Chambon during the Occupation. The precise, sixty-year-old daughter had told me her version of Theis's return to Le Chambon from a detention camp. She told me that when Theis and the other leader of Le Chambon stepped off the train, there were the villagers, waiting, with an open path through them. There was absolute silence at the railroad station. Not a word was said; there was not even a cry from a child and no shuffling of feet, she said. She told me that the two returning leaders had walked through the open path, and the villagers had silently closed behind them and followed them away from the station.

Theis, when he heard this, was surprised. "I remember no silence. Perhaps if Trocmé were here, he would remember it," he said.

"Why," the younger Marion insisted in her quick, factual way, "Monsieur Theis, they were silent, and silent for a very good reason. The Gestapo was there, and so were some collaborators from Vichy. They were looking for an excuse to arrest you again. Didn't you notice?"

"Oh, perhaps I did notice," Theis murmured, "and I have forgotten. But I am not sure. I am not sure at all. And André Trocmé is dead." The story of Le Chambon was being swept out of human memory.

The Baal Shem-Tov, the founder of Hasidism, the Jewish movement that finds God in good and evil—in everything—once said, "In remembrance resides the secret of redemption." This saying appears at the top of a citation from the state of Israel, a citation that attests to the fact that André Trocmé, the spiritual leader of the village of Le Chambon, "at peril of his life, saved Jews during the epoch of extermination." This citation accompanied the Medal of Righteousness that Israel awarded to Trocmé. If the secret of redemption lies in remembering, it is lost in forgetting. After more than thirty years, the story of Le Chambon lay in less than a dozen little-known pages, most of them rather vague and inaccurate. The whole story could be found only in the memories of a few people who were now old and sick.

Walking with Pastor Theis that evening in January of 1976, I was afraid that I did not have the gifts needed to uncover the "secret of redemption" that lies in the story of how Trocmé and the people of Le Chambon saved human lives at the peril of their own. I knew that I could not tell the story as thoroughly as a careful historian might tell it; I was neither trained nor inclined to report every detail I could find. Nor was I religious enough to be certain of exactly what the Baal Shem-Tov meant by "redemption." But I believed that here in Le Chambon goodness had happened, and I had come to this village on a high plateau in southern France in order to understand that goodness face-to-face.

Having lived the life I have, I have never doubted that evil, in the sense of grievous harmdoing, is possible. It has happened, and it has happened to me. In New Lenox, Illinois, when I was seven years old, two blond, older boys smashed and bloodied my face because I was a Jew. As an artilleryman in World War II, I passed the bodies of human beings lying beside the road, chopped by our artillery into foot-long chunks, like fresh meat in a butcher shop. For more than six months I was in a gun crew that shot 155-millimeter shells into German troops, and since that time there has been a high-pitched humming in my head—every moment of my waking life—to remind me that I have killed human beings. If only such things were possible, then life was too heavy a burden for me. The lies I would have to tell my children in order to raise them in hope—which children need the way plants need sunlight—would make the burden unbearable.

I am a student and a teacher of good and evil, but for me ethics, like the rest of philosophy, is not a scientific, impersonal matter. It is by and about persons, much as it was for Socrates. Being personal, it must not be ashamed to express personal passions, the way a strict scientist might be ashamed to express them in a laboratory report. My own passion was a yearning for realistic hope. I wanted to believe that the examined life was more precious than this Hell I had dug for myself in studying evil.

But it was not easy to find out what had happened in this village. From the point of view of the history of nations, something very small had happened here. The story of Le Chambon lacked the glamour, the wingspread of other wartime events. Victories and defeats of nations are written large in men's minds because the lives and fortunes of so many human beings depend upon them. World War II, between the Axis and the Allies, was a public phenomenon; military, journalistic, and governmental reports made it abundantly available to the public. It impressed itself powerfully and deeply upon the minds of mankind, both during and after the war. The metaphors that describe it have a flamboyant cast: the war itself was a "world war," with many "heroes"; there were "theaters of war," and soldiers who participated in major "campaigns" received "battle stars."

No such language applies to what happened in Le Chambon. In fact, words like "war" are inappropriate to describe it, and so are words like "theater." While the story of Le Chambon was unfolding, it was being recorded nowhere. What was happening was clandestine because the people of Le Chambon had no military power comparable to that of the Nazis occupying France, or comparable to that of the Vichy government of France, which was collaborating with the Nazi conquerors. If they had tried to confront their opponents publicly, there would have been no contest, only immediate and total defeat. Secrecy, not military power, was their weapon.

The struggle in Le Chambon began and ended in the privacy of people's homes. Decisions that were turning points in that struggle took place in kitchens, and not with male leaders as the only decision-makers, but often with women centrally involved. A kitchen is a private, intimate place; in it there are no uniforms, no buttons or badges symbolizing public duty or public support. In the kitchen of a modest home only a few people are involved. In Le Chambon only the lives of a few thousand people were changed, compared to the scores of millions of human lives directly affected by the larger events of World War II.

The "kitchen struggle" of Le Chambon resembles rather closely a certain kind of conflict that grew more and more wide-spread as the years of the Occupation passed. Guerrilla action, clandestine, violent resistance to the German occupants, was as much a part of the history of that Occupation as the story of that little commune. Secrecy was as vital to guerrilla warfare as it was to the resistance of the people of Le Chambon, and so was a minimum of permanent records. In both cases military weakness dictated that there be few records and much secrecy.

But the kitchen struggles differed greatly from the bush battles (the *Maquis,* the name given to the wing of the armed resistance which had no direct connections with de Gaulle's Secret Army, refers to *le maquis,* the low, prickly bushes that grow on dry, hilly land). The guerrillas were fighting for the liberation of their country. Some of them received their orders from de Gaulle's exiled French government (Free France), and others owed their allegiance to the Soviet Union; still others had no particular political allegiance, but they were all parts of military units. Especially in time of war such units are primarily concerned with achieving by violent means a victory over the enemies of those units. Their primary duty is not to save lives but to save the life of some public entity; and especially in time of war they cherish heroism—living and dying gloriously for a public cause—more than they cherish compassion. The consciences of individuals in military units tend to be in lock-step with the self-interest of the units. In fact, for the bush warriors as for the uniformed warriors, public duty took precedence over personal conscience.

But the people of Le Chambon whom Pastor André Trocmé led into a quiet struggle against Vichy and the Nazis were not fighting for the liberation of their country or their village. They felt little loyalty to governments. Their actions did not serve the self-interest of the little commune of Le Chambon-sur-Lignon in the department of Haute-Loire, southern France. On the contrary, those actions flew in the face of that self-interest: by resisting a power far greater than their own they put their village in grave danger of massacre, especially in the last two years of the Occupation, when the Germans were growing desperate. Under the guidance of a spiritual leader they were trying to act in accord with their consciences in the very middle of a bloody, hate-filled war.

And what this meant for them was nonviolence. Following their consciences meant refusing to hate or kill any human being. And in this lies their deepest difference from the other aspects of World War II. Human life was too precious to them to be taken for any reason, glorious and vast though that reason might be. Their consciences told them to save as many lives as they could, even if doing this meant endangering the lives of all the villagers; and they obeyed their consciences.

But acts of conscience are not important news, especially while a war is going on. Only actions directly related to the national self-interest receive a measure of fame then. And this is why the Partisan Sharpshooters of the left and the Secret Army of General Charles de Gaulle were nationally though secretly revered during the Occupation, and were praised to the skies afterward. This is also why the armed resistance produced heroes like General de Gaulle himself, and the passionately beloved coordinator of the armed French Resistance, Jean Moulin.

There are no such nationally known names in the story of Le Chambon. When France was liberated, there were no triumphal marches for André Trocmé and his villagers through the streets of Paris or Marseilles. And this was as it should have been: they had not contributed directly to saving the life of the French nation. They were not so much French patriots as they were conscientious human beings.

In fact, as soon as the European war ended, the commune of Le

Chambon went into a sharp decline. Almost all of those who had come to her for help left when the trouble ceased. The international, pacifist school that Trocmé and Theis had founded there before the war, the school that had sheltered so many refugees during the first four years of the decade, almost perished for lack of money and interest. Soon after the liberation of France, Trocmé had to go to America, hat in hand, in his only decent suit and his only pair of leather shoes, in hopes of finding money and workers to keep the school—and Le Chambon—alive. And of course what he found in America was complete ignorance of what his Chambonnais had done. There are many friends for the rescuers of nation, but there do not seem to be many sympathizers for the rescuers of a few thousand desperate human beings.

It is plain that the story of the struggle of Le Chambon is of no special political or military interest. But it is of ethical interest. The word *ethics* can be traced to the Greek word for character, an individual person's way of feeling, thinking, and acting. Ethics is concerned with praising some sorts of character and blaming other sorts. In that region of ethics concerned with matters of life and death (as against, for example, sexual or professional conduct), a person who destroys human life is blamed for doing this unless that destruction can be excused or justified in some way; and in life-and-death ethics a person who avoids or prevents the destruction of human life is praised for doing this unless his deed can be shown to be destructive of human life. In life-and-death ethics the person we blame is often called "evil," and the person we praise is often called "good," though we may use such milder terms of praise and blame as "bad" and "decent."

It is this important way of judging human character that helps us to understand what happened in Le Chambon in terms close to the story itself. In this village the characters of individuals were of immense importance, and most of these individuals were dedicated to protecting human lives instead of destroying them. It is as persons rather than as parts of some public entity that we shall find most of them praiseworthy according to a life-and-death ethic. In their conflict with those bent on diminishing or destroying human lives lies the story of Le Chambon.

What had brought me to this ice-covered village high in the mountains of France more than thirty years after the Liberation was the need to understand the story that connected these two kinds of individuals with each other and with their time. I needed this understanding in order to redeem myself—and possibly others—from the coercion of despair.

During the first few moments after Theis and I left the boardinghouse of the Marions, we walked separately through the snowy mountain wind. Two or three times Theis staggered and almost fell. Even I, with my younger reflexes, slipped once or twice. But when he reached out and intertwined his right arm with my left, suddenly the warmth of his thin body and the firmness of our intertwined arms created a new being moving upon four firm legs. Now we were stable, even though the icy road was still there, and even though the broom were still whirling their long evergreen fingers. The world was still cold, confusing, and dangerous. But we were close to each other, parts of a new whole, and we felt suddenly surefooted.

The day before this walk we had been in the home of the Chazots, who had kept a Jewish family in their home for most of the four years of the Occupation, even though their house is on a road that was much frequented by Vichy and German troops. During our conversation Theis had said suddenly, *"Oh! Que c'est difficile d'etre seul!"* At that moment his phrase, "Ah, but it is hard to be alone!" had only the pathos of bereavement in it: Mildred Theis was dead. But now, when I have thought long about our walk and have learned the story of Le Chambon, the phrase means, "Ah, but it is a joy to be together, always joyously good!"

Le Chambon-sur-Lignon
Haute-Loire
France

PART ONE

Persons

1

The Arrest of the Leaders

1.

At seven o'clock in the evening of February 13, 1943, an official black automobile stopped not far from the Protestant presbytery of Le Chambon. Automobiles—after three years of German occupation—were very rare in those days. But the car was not noticed much, partly because the street it stopped in was a secluded one, and partly because in the last few months there had been an increase of official, especially police, activity in the village. The Germans were learning a new feeling, the feeling of danger. A few months before, the British and the Americans under the command of General Dwight D. Eisenhower had landed on the beaches of Morocco and Algeria, and Nazi North Africa was crumbling. To protect conquered France from this new threat to the south, the Germans had moved their troops, their secret police, and their administrators across the demarcation line at the Loire and into southern France, where Le Chambon stood. Hitherto they had been content to stay above the Loire and on the vulnerable west coast of France, and had been content to give the French the illusion of governing themselves in the south, but now they had a southern flank to cover in Fortress Europe. Moreover, the Russians at Stalingrad had checked the momentum of the great Nazi military machine, when just a few months before that machine had seemed irresistible. And so the Germans were disturbed, and were showing their disturbance by sending their instruments, the police of Marshal Philippe Pétain's Vichy government, into the quiet regions of southern France in order to threaten or arrest anyone who might abet their now dangerous enemies.

Out of the car stepped two uniformed Vichy French policemen: Major Silvani, the chief of police of the department of Haute-Loire, and a lieutenant.

The winter wind, *la burle,* was twisting and piling snow around the gray, fifteenth-century presbytery when they knocked on the door. It was a very dark night, without moon or stars, but the accumulated snow vaguely lit up the long dining room in which the minister's wife, Magda Trocmé, was working. Through its three windows cut in the thick granite walls, the dim, reflected light made barely visible the sycamore and oak paneling on the ceiling and on the walls.

Pastor André Trocmé was away when they arrived. This was a day of "visits," when he dragged his big-boned body across the village and countryside into the kitchens of his parishioners. On snowy days like this the visits were especially painful for him because the snow was often hip-deep and his back was a girdle of pain as he trudged forward. Once, in far better weather, a farmer had found him stretched out on a hillock, paralyzed with pain because of all that walking to the far-flung homes of his parish. The farmer had brought him back to the presbytery in his milk cart.

It was these visits, more than any other part of his ministry, that made him the trusted and beloved leader of Le Chambon. His sermons were powerful, but these visits brought him into the center of the daily life of his people. In the intimacy of a home his excitement and his drivingly penetrating blue eyes drew his parishioners to him.

Some of the houses he was visiting were the homes of the *responsables,* the leaders of the thirteen youth groups Trocmé had established in the parish in order to study the Bible in intimate groups. When the Germans conquered France these groups and their leaders became the communications network and the moving spirits of a village committed to the cause of saving terrified foreigners from their persecutors. During the Occupation Trocmé gave most of his instruction—on the Bible as well as on resistance—to these leaders as they sat in his somber office in the presbytery; but it was necessary that he visit them individually as well. He had to be the only person in Le Chambon who knew about the entire operation; he had to see to it that the groups operated as independently of each other as possible, so that if one leader were tortured, he could not reveal enough to destroy the whole rescue machine. This meant that each leader had a heavy responsibility and had to make swift, intelligent decisions on his own when Trocmé was not available. But it also meant that the *responsables,* many of them young people, often sought out the clarity of mind and the unswerving commitment of their pastor in order to do their essentially lonely jobs. And so André Trocmé made their homes part of his itinerary of visits more often than he made the homes of others, except for the sick.

Now that German troops and secret police were in the south, it was more important than ever that the young leaders make no mistake and that they never waver a hairbreadth in their commitment to saving the lives of Jewish refugees. Before the Germans came, Vichy had been known to be lax as far as reprisals were concerned, but the Germans were capable of massacring a village that dared to resist their will—and Le Chambon had the reputation of being a "nest of Jews in Protestant country." The times demanded close communication between Trocmé and his *responsables.*

This is why Trocmé was away when Major Silvani and his lieutenant knocked at the presbytery door. The wife of the minister was surprised to hear a knock at night. Refugees usually came during the daytime, after the one o'clock train; they usually knocked; but at night only villagers came to the presbytery, and they knew they were expected to walk right in. At night there was usually so much happening in the house, with the Jewish refugee cook sadly answering complaints about burning food (she had no butter or lard) with her heavy German accent, another refugee doing noisy cabinet work in the dining room or the attic, other refugees following Magda Trocmé around the house as she performed her duties, the four turbulent Trocmé children, and the people of the parish coming and going, that one could understand why a knock was likely to be unheard.

Somehow Magda Trocmé heard the knock, and having warned the man from Berlin to go up to the attic, and the woman from Karlsruhe to go to the cellar (the cook and the cabinetmaker had the nervous habit of popping their heads into a room and asking, in the middle of an interview, if the coast was clear), she went to the heavy old door and opened it. The dark, funereal-voiced Silvani stood in the doorway with his lieutenant, apologetically but with dignity. Rolling his *r*'s with his Corsican accent, he asked where her husband was. She said that she did not know, and asked them to come in and wait for him in his office on the other side of the dining room. She led them there, turned on the light, and closed the door slowly behind her. The moment she and her husband had been expecting for many months had come.

After a couple of hours, her big husband pounded into the house, strode across the dining room, and leaped at the door to his office in order to put some papers there. He moved like that, by bursts and leaps, despite his bad back. In a few minutes, he came back into the dining room and said, "Magda, I am arrested."

She was an impulsive Italian, and she cried, "Oh, André! What about the suitcase with all those warm clothes in it, the clothes I packed for you months ago? We've been using them! The suitcase is empty." And there they stood, facing each other for a moment in dismay.

André Pascal Trocmé was born in 1901, and being almost as old as his century, he was in his full vigor during the early 1940s when he was head of the Protestant parish of Le Chambon and leader of nonviolent resistance in his part of France. On his father's side, he was a descendant of a long line of Huguenots (one of his ancestors is rumored in the family to have been one of the original followers of John Calvin in the sixteenth century); on his mother's side, he was descended from Germans. He was more than six feet tall, fair-skinned, and wore glasses over his piercing, light-blue eyes. His energies were immense, whether they took the form of sometimes embarrassing affection or of equally embarrassing anger.

His wife was born Magda Grilli in Florence, Italy, also in 1901. She was a descendant of one of the leaders of the Russian Decembrists who had tried to free the serfs long before the Revolution of 1917. During the Occupation she too was in her full vigor. She wore her brown hair in braids upon her head, and her great energies—fully as great as her husband's—took the form of unending, compulsively scrupulous work on behalf of any person —child or adult—who needed warmth and food.

There they stood, looking into each other's eyes: two excitable, aggressive creatures accustomed to having titanic arguments with each other and accustomed to kissing each other tenderly at the end of every day in which an argument had occurred. In less dangerous times they might have burst out laughing at themselves and at each other about getting concerned over a valise. But the prison suitcase, which they kept in their bedroom closet upstairs, was a witness to the precarious life they were leading.

They were never sure that the next day would not bring arrest and possibly deportation to a concentration camp in Central Europe.

She went straight to the basement to reassure "Madame Berthe" that the police were there for the pastor, not for her. Then she rushed up to their bedroom to pack the valise with the clothes that might make the difference between life and death for her husband. André Trocmé went up to the attic to explain to "Monsieur Colin" the mission of the police, and to warn him not to open any doors to hear what was being said.

The pastor and his wife had been invited to dinner by parishioners, the Giberts. After Trocmé had been home for about a half hour, the daughter of the Giberts came to remind them of the dinner engagement (they often forgot such invitations). When she entered the dining room, she saw the police arresting her pastor. She turned and rushed out of the house. In a little while, not only her parents but many of the other villagers in this tightly knit Huguenot community knew that André Trocmé had been arrested.

After the pastor and his wife rejoined the police in the dining room, Magda Trocmé invited the two policemen to have dinner with them. Later, friends would say to her, "How could you bring yourself to sit down to eat with these men who were there to take your husband away, perhaps to his death? How could you be so forgiving, so decent to them?"

To such questions she always gave the same answer: "What are you talking about? It was dinnertime; they were standing in my way; we were all hungry. The food was ready. What do you mean by such foolish words as 'forgiving' and 'decent'?"

Magda Trocmé was not the only citizen of Le Chambon who scoffed at words that express moral praise. In almost every interview I had with a Chambonnais or a Chambonnaise there came a moment when he or she pulled back from me but looked firmly into my eyes and said, "How can you call us 'good'? We were doing what had to be done. Who else could help them? And what has all this to do with goodness? Things had to be done, that's all, and we happened to be there to do them. You must understand that it was the most natural thing in the world to help these people."

Often moral praise is an interpretation, a grid laid upon the facts by an outsider's hand. An outsider may see goodness or decency as something "in" an action, as an integral part of, say, Magda's invitation to the police to join them at the big dining table, but the doer of the deed, who often acts on the spur of the moment, sees nothing "moral" or "ethical" in that deed. For Magda Trocmé and most of the other people of Le Chambon, words of moral praise are like a slightly uncomfortable wreath laid upon a head by a kind but alien hand.

They sat down to eat, with the four Trocmé children watching Silvani with great, staring eyes. When Magda Trocmé offered him food, he told her that he did not have the heart to eat, and his face hung down almost to the plate in his misery. André Trocmé, as was his custom, ate heartily, and even more than usual, just in case imprisonment was as bad as he had heard it was.

In the course of the meal, he asked the chief of police why he was being arrested. Silvani answered, "I don't know. I know nothing. I can say nothing"; and the dignified, black-haired policeman was on the verge of weeping.

Upstairs, "Monsieur Colin," the cabinetmaker (whose real name was Kohn, but whose identity card had been falsified to conceal his Jewish name), was frantic with impatience, eager to emerge from his hiding place in the attic. But the four refugee Jews in the house—Kohn, Madame Grünhut ("Madame Berthe," the cook), and two students from the Cévenol School—managed to keep each other still and remained concealed.

Soon after the meal began, parishioners started coming into the presbytery to say their good-byes to their pastor. Most of them were crying, and some of them had whispered criticisms to make of the police who were sitting there, heads bowed, at the table. One of them, during a warm but shy embrace, whispered into the minister's ear, *"Quel mal avez-vous fait?"* ("What harm have *you* done?") The French word *mal* means both "evil" and "harm," and the rhetorical question meant: "We know that the *laws* of Vichy and the Nazis have been broken by you and by us, but we have done no evil because we have done no harm to our fellowman; in fact, we have tried to help those whom the law was designed to hurt."

The people of Le Chambon knew about the false identity cards; they knew that sheltering foreign refugees and not registering them under their true names was in violation of the laws of France. But they also knew that sometimes—and this was one of those times—obeying the law meant doing evil, doing harm. It all lay in that one little French word *mal:* evil was harm, nothing more, nothing less.

As they embraced Trocmé, they put in his hand precious packets. And after each group left, the minister brought these to the dining table. There were candles, which were all but unobtainable at this time, there were long-forgotten luxuries like sardines and chocolate biscuits, and there was one big sausage. There were warm stockings, and on top of the heap Trocmé had made on the dining table there was a roll of toilet paper, which was a precious gift in those poverty-stricken days.

At first the two officers watched all this with wide-open, amazed eyes. But as it went on and on, they seemed to crumple, and Silvani said, "I have never seen such a farewell, never." And he sat there, his head bowed, weeping over his untouched food. So distraught was he that never once did he pose the question that the police usually asked in the home of a Chambonnais: "Are there any refugees here?" He was following his orders softly, minimally; he was simply arresting the pastor of the village of Le Chambon.

Someone brought in a long, thin votive candle and handed it to Magda Trocmé. After the giver had left, she said, "But where will he get matches to light this?" Silvani walked up to her with a box of matches in his hand and said, "Please take these." Then he added, "I'll tell this story later. Yes I'll keep it alive."

Later, in his report, he would describe the people of Le Chambon as "full of love." He had entered the village expecting trouble, had cut off all telephone and telegraph communication between the village and the rest of the world, and had surrounded it with armed police squads.

Now the tears in his black eyes expressed moral praise and moral dispraise more powerfully than any words could do. And every move the people of Le Chambon made in presenting their gifts, in kissing their pastor good-bye in those dangerous times, and in scowling at the policemen, expressed their moral praise and their moral dispraise. Though Magda and André Trocmé themselves were innocent of any thoughts about their own decency, they were surrounded by people who were having such thoughts, surrounded by the airy forces of ethical judgment—and all those forces were in agreement with each other. They all saw André Trocmé as a good man, and they all saw harm as evil. Ethics is an interpretation men lay upon the facts, but men can agree on that interpretation.

When André Trocmé said good-bye to his children, he did not know if he would ever see them again. The Maquis around Le Chambon were getting more numerous and more violent, and so were the German troops and the Gestapo. Those days were full of disappearances and sudden death. But he was calm, almost joyous, not only in order to keep his family from knowing terror but because for a long time he had been hoping for a test, a hard test. His warmth, the speed of his intelligence, the vigor of his pain-wracked body that could work efficiently with only a few hours of sleep, *and* his luck had kept death away from him and his family and had brought admiration and love to them. Moreover, he had been efficient: he had created a rescue machine made of poor people who had enough problems of their own to keep them fully occupied. But he did not know whether he could be helpful under great pain and under the immediate threat of death. Now he would find out. He was like an eager chemist watching the results of one last acid test that would tell him if the substance before him was, indeed, gold. The substance before him was himself.

Magda gave him three items: his freshly packed valise, a bulky package containing the gifts, and a pair of wooden shoes. In those days the people of Le Chambon wore wooden shoes because they could not afford leather, and also because it was the custom, especially for the peasants on the farms with their stony roads and mucky barnyards. Rumors about the internment camps told of their filth, but for both Magda and André Trocmé those shoes were mainly a palpable sign of his oneness with the people of Le Chambon, a sign of his involvement in their everyday lives.

The three men stepped out into the Rue de la Grande Fontaine, which runs along the north wall of the presbytery. The narrow, medieval street was dark, and *la burle* was blowing thin snow around the broken, ice-covered stones in the road. On both sides of the crooked street, their feet in the thin snow, villagers were lined up, looking fixedly at Trocmé as he walked between the two policemen. Standing among the villagers were refugees from Central Europe, and students and teachers from the Céve-nol School. There were also Darbystes, rival Protestants who did not believe in the necessity of having ministers or churches.

As the three walked west down the street toward the high road that led to the village square, the bystanders began to sing the old Lutheran hymn "A Mighty Fortress Is Our God." A woman named Stekler, sister of a half-Jew who had been arrested and released by the Vichy police, started the singing. The calm, deeply rooted song surrounded the three men, while the villagers closed behind them, and the *clop-clop* of their wooden shoes, muffled a little by the thin snow, followed them up the street.

The singing of Luther's hymn embodied the Protestantism of the village. In this town of about three thousand residents there were only a few more than a hundred Catholics. But throughout France considerably less than 1 percent of the population is Prot-estant. That walk up the Rue de la Grande Fontaine and the

whole story of Le Chambon in the first four years of the 1940s cannot be comprehended, in their depths, without some understanding of the history of what was now an infinitesimal minority in France, the Protestants.

In the sixteenth century, when Protestantism came to France in the persons of Lutherans or Huguenots, the Protestants of France entered into a long period of suffering that would extend across three centuries into the French Revolution of the late eighteenth century. The Saint Bartholomew's Day Massacre of 1572 was only one of the many bloody persecutions that the Huguenots suffered. For three centuries, with only a few tiny cases of tolerance, those who were found to be loyal to the Protestant Reformation of Christianity were often stripped of their property, their liberty, and even their lives. The king of France had in the eighteenth century great galleys which were manned by men whose only crime was that of being Protestant. For this crime they slaved away their short, bitter lives. And there was the Tower of Constance near Marseilles, where women convicted of the same crime were left to die of starvation, of cold, of heat, and of despair. Until the French Revolution, France was almost always *le désert* (a wasteland) for her Protestants. Their temples (and even this word originated as a term of derision suggesting pagan rites) were razed, so that worship had to be conducted in darkened homes or in secluded fields and woods. The Protestants had learned to expect all of this from the law of France: *le mal* (evil, harm).

During these years, the French *pasteurs,* the Protestant shepherds of endangered flocks, were often living, visible counterparts of a persecuted Jesus. Le Chambon was an old village when Protestantism came to it in the first half of the sixteenth century. During hundreds of years of persecution, her pastors and her people were arrested by the dragoons of the king and then hanged or burned either in Le Chambon itself or in Montpellier to the south.

And it was a village with a stable population, a village whose history was not buried in unread books or forgetfulness. After the war, Roger Darcissac, as historian of the commune, made a demographic study of Le Chambon-sur-Lignon. He found that in the three hundred years following the Protestant Reformation in Europe, the population of the town had increased in number by only a thousand. The steepest, most dramatic change in the size of the population came with the virtual destruction of Protestant rights in France when the tolerant Edict of Nantes was revoked in 1685. At that date almost the entire thousand newcomers to Le Chambon came to the village for protection. The village assimilated the newcomers, and the population became stable again until the 1930s, when Theis and Trocmé rejuvenated the village by starting the Cévenol School.

For such a stable, tightly knit population, history was personal and important; the story of the Huguenots was alive in the minds of the people of Le Chambon. In the course of the Occupation, Darcissac put out a book of songs for young people. In it there was a *complainte,* a song praising and mourning a pastor of Le Chambon who had been arrested during *le désert* and had been hanged in Montpellier. The book was illustrated, and one of the pages displayed a picture of one of the royal dragoons astride his horse, which was standing upon sand. Sand is for the Protestants of France a symbol of the desert that France was for them before the Revolution. Instead of drawing sand particles under the horse's hooves, Darcissac had the illustrator spell out certain messages in Morse code, dots and dashes, messages of indignation and of consolation for the people of Le Chambon and of France.

Solidarity and resistance not only in the face of persecution but against the national establishment of France, the law of the land, combined with devotion to the pastors who have maintained that solidarity and have led that resistance, are central to the spirit of Protestantism in France. These feelings, refreshed and personalized by Trocmé's painfully achieved "visits" and given a public form by Trocmé's sermons in the temple against harmdoing, were a part of that demonstration that dark night in the Rue de la Grande Fontaine.

When they came to the main square of the village, the three men entered a half-track, with Silvani sitting in front next to the driver, leaving the lieutenant in the back seat with Trocmé. In his Corsican accent with its heavily rolled *r*'s, the major turned to Trocmé and announced, "We are going to find your colleague, Édouard Theis, and then we shall pick up the director of the public school, Roger Darcissac."

Édouard Theis lived near the school he directed, on the outskirts of the town. Because of his location and his work, he was not so deeply involved in the everyday affairs of the village as Trocmé. He was a half-time minister under Trocmé and sometimes preached in the temple, but his main task was that of administering the Cévenol School with the help of a codirector. Theis was born in 1899, and before Trocmé brought him to Le Chambon he had been a missionary and a teacher in Africa. He had been a friend of Trocmé in the faculty of theology at the University of Paris, and was a student and teacher of ancient and modern languages. He was as tall as Trocmé, but more massive, with a wide face and a large, almost pointed, aquiline nose. Though he was a student and teacher of languages, he spoke little and was almost as quiet as his wife, Mildred, who had been born and raised in Ohio. What the two shared—apart from their eight daughters—was commitment. In their way—so different in its shyness from the way of the Trocmés—they were unshakably committed to obeying the Sermon on the Mount, to strengthening and expanding the Cévenol School, and to implementing the ever-fresh ideas of André Trocmé.

Though the Cévenol School had many refugees in it as teachers and students, and though Theis had helped all of them get false identity and ration cards, he was not always the first to hear about the new arrivals or about new plans concerning the refugees. Still, because of his sermons in the temple and because of his protection of the refugees, he had reason to fear being arrested. He was one of the leaders of this recalcitrant commune.

Silvani seated him in the back seat with Trocmé and the lieutenant, and they went on to find the third leader of the village, Roger Darcissac. On the way, Trocmé was afraid for Darcissac: he was the photographer of the counterfeit cards, and so, directly involved with the crimes Theis and Trocmé had been committing.

With his big shock of dark hair and his frequent changes of mood, he was the most worldly, the most sensual of the three leaders. Though he held an important official position—he was head of the public Boys' School situated across the road from the Protestant temple of Le Chambon—he was the most volatile and unpredictable of the three leaders. He was—and is—like one of the boys he taught, eager to laugh, yearning to try new foods, and ready to be angry in defense of his ideals. The deepest force that bound him to the two ministers was his devotion to French Protestantism. Not only did he believe in Trocmé and Theis; he knew in detail and cared deeply about the long history of persecution that the Huguenots had suffered in France.

By the time the police arrived at his house, he had fled. With quiet dignity, Silvani convinced Madame Darcissac that her husband's flight would endanger not only himself but his two colleagues, Theis and Trocmé, and possibly the whole village. She sought him out, and after a short talk with him, she walked with him to the half-track, where Silvani put him under arrest. His slender, long face was pinched with fear, a frantic fear quite unlike Theis's fear, which was like the man, stolid.

All this time, Magda Trocmé was sitting in a police car that had been following the half-track: she was going to be as near as possible to her husband while he was within the limits of Le Chambon, and she had made this painfully clear to Silvani's lieutenant. Her car followed them until they reached the village limits. Then the lieutenant took her back to the presbytery. The leaders went on to the nearby town of Tence for registration in the police station there, and then they started out for the city of

Le Puy. On the way to Le Puy, Silvani turned to Theis and Trocmé (Darcissac was in another vehicle) and tried to reassure them. "Things will change. You'll not be there long." Knowing the times, the two men were not reassured.

As the chief police representative of Vichy in the department of Haute-Loire, this sentimental yet opportunistic man had the difficult task of enforcing laws that some of the most influential leaders in his department—like Trocmé himself—despised. He did his job *doucement* (mildly and compassionately). In 1955, more than a decade after the Liberation, Trocmé would meet Silvani again, now as a colonel in Algeria. After the war Trocmé wrote a letter that restored Silvani to favor with the French government and the angry people of France, despite the high office he had held in the Vichy government. When Trocmé saw him in Algeria in 1955, he was once again a "patriot," an enforcer and defender of the law of the land, no matter what that law might be.

2.

The internment or concentration camp of Saint-Paul d'Eyjeaux is near the city of Limoges on the great Plateau du Velay. It is situated on the northwest corner of that plateau, and Le Chambon-sur-Lignon is diagonally opposite it, on the southeast corner of the same plateau. Now traveling by train, and having left Silvani behind, the three men found themselves making one more stop, this time in Limoges, before entering the camp. The police chief of Limoges was convinced that anybody who had been arrested was not only guilty of a crime but beneath contempt. Facing him in a police station, they heard him mutter as he looked at their orders, "Pastors! A teacher!" Then he whistled in mock wonder. "Where will evil hide itself nowadays? Come on, out with it! What have you done? Confess! The black market? Swindling, maybe?"

The prisoners answered, "We don't know the charge. Maybe

they've arrested us because we have been trying to save Jews from being deported."

The captain's fury exploded. "What? Jews? Oh—that's lovely. Now that doesn't surprise me. You're part of their conspiracy, eh? We all know that they're the ones who have brought France down into the abyss. Well, you're going to pay for this. You're going to pay for all the harm you've done to the marshal!"

This was a moment Trocmé would never forget. In fact, his overnight stay in the police station in Limoges changed his view of mankind. He discovered people like the captain—patriotic, sincere, but above all, severely *limited.* These people were capable of repeating hate-ridden clichés without any concern for evidence or for the pain of others. Before he entered that police station in Limoges, he thought the world was a scene where two forces were struggling for power: God and the Devil. From then on, he knew that there was a third force seeking hegemony over this world: stupidity. God, the Devil, and halfwits of mind and heart were all struggling with each other to take over the reins.

Le Chambon had been spared the cretins. In the south of France, people called the village the "republic of Le Chambon" because somehow it managed to remain a world of its own, an impregnable fortress in a murderous world, a place that could not be made a party to the compromises and murders of the France around it. This moral isolation had kept out the cretins, or at least had quieted them and assimilated them so that Trocmé had not seen them parading their cretinism. Now and for the rest of his life he knew that there were some people—indeed, many people —who did not *realize* what suspicion and hatred were doing to their own minds and to their victims.

The next day was Sunday, and instead of being allowed to go to services in Limoges (Trocmé and Theis had friends in Limoges who were pastors), they were put in a sealed bus with barred windows and taken down into the valley where the camp stood. A typical concentration camp, it consisted of low, gray wooden barracks surrounded by two ranks of high, barbed-wire fences, dominated by watchtowers in which stood guards armed with machine guns. Outside the barbed wire stood the administration building of the camp, where prisoners were registered, their fingerprints and photographs taken, and numbers given them to replace their names. Personal possessions were taken from them, and noses were measured to ascertain whether or not they were Jewish. The French police were, in their national policy if not in their local behavior, as anti-Semitic as the Nazis could have wished them to be. As a matter of fact, Vichy had a definition of the word *Jew* that covered and condemned more people for the crime of being Jewish than the current Nazi definition. With the help of the cretins, Vichy was going out of its way to please the masters of France.

Seeing the camp, André Trocmé felt fear pinch his heart for the first time. In this, the third year of the Occupation, he knew about the death camps of Central Europe. The starvation, the torture, the mass killings were clear in his mind because refugees and Aryan German resisters stopping in Le Chambon (one of them bearing the glorious name of Goethe and looking as noble as Goethe himself) had described the camps in detail at the dining table of the presbytery. Now Trocmé found himself thinking, Well, this is the antechamber to a German camp. We are on our way to death.

When they passed into the "no-man's-land" between the two barbed-wire fences, he saw confirmation of his fears. The three Chambonnais found themselves facing about thirty prisoners (out of what proved to be about five hundred in the camp) lined up to watch them. They were wearing ragged army coats, and their faces were yellow and slack.

"Who are you?" one of the walking corpses asked.

"Two preachers and a teacher," the trio answered.

The inmates guffawed: "Pastors! That's all we needed! We've got a priest and a rabbi, and plenty of teachers, and now we've got pastors too!"

The welcome was openhearted, even with the vein of irony, and the three found themselves in a situation where gaiety and a happy madness reigned. Cadaverous as they were—their yellow complexion may have come from the only staple of their diet, rutabagas—the prisoners' camaraderie was magnificent and lifted the three men's hearts.

That first evening in the camp, Trocmé, joking with the others, opened up the roll of toilet paper in order to share it with them. On the outer sheets he found written in pencil verses of consolation from the Bible. He stopped laughing, and so did his new friends. Magda believes the Darbystes, who knew the Bible by heart, might have written them; Trocmé himself believed it was a member of his parish. But whoever wrote them, those verses reminded Trocmé that he was still a part of Le Chambon.

3.

In the course of a few days the Chambonnais learned much about their fellow inmates. They learned that most of them were the leaders of the most important Communist cells in southwest France; some of these leaders had been interned here since the Hitler-Stalin pact of 1939. Some of them were Catholics who had opposed Vichy's dictatorial and anti-Semitic policies. And there was one nonbelieving Protestant whose only mark of distinction was that he attached himself to the Chambonnais from the very beginning, only out of greed for the gifts they were getting from the village.

At first the three nonviolent Protestants were severely criticized by some of these people. Some thought that they were, or at least might well be, *moutons* (black sheep) who had been placed there to betray the most active resisters by passing information to Vichy about their hopes and hatreds and plans. But these made a small group, or at least a quiet one. The group who openly and regularly attacked the Protestants was the Communists, who were angry and bitter at their nonviolence. "You refuse to kill?" they would say. "Why, in war—and we are in a state of war with Vichy and Germany—that's aiding and abetting the enemy! You're peddling the same old opiate of the people that has kept the masses from moving forward to social justice!"

But it did not take Trocmé long to show them that they and their people in Le Chambon were as vigorous and daring in their resistance to Vichy and Germany as the most aggressive inmates in the camp. He had always disliked intensely the connotations of the term *pacifist,* with its suggestions of passivity and even retreat, and his few remarks about the activities in the village of Le Chambon swiftly persuaded most of them that, in their own way, by their own principles, the people of the village were doing the best they could against the powers dominating Europe and threatening to dominate the world.

One evening in the course of their first weeks in the camp, Trocmé and twenty-nine other prisoners were listening to a BBC broadcast emanating from a radio concealed in a jar. Suddenly the announcer stopped the quiet flow of information with an announcement: the battle of Stalingrad was over; the Germans had suffered the most terrible defeat in the history of the Third Reich. All of the thirty men in the barracks room burst into cheers.

But the same outward signs concealed basic inward differences among them. Those patriotic followers of de Gaulle, the members of his Secret Army, were full of joy at the prospect of their country's liberation after three years of deprivation, humiliation, and death at the hands of the Germans. To them, this was the beginning of France's rebirth. The Communists, most of them Partisan Sharpshooters, cheered for the victory of Russia and of Communism. They saw this not as a national matter, but as one involving all the downtrodden peoples of the world. At last institutionalized cruelty, capitalistic Fascism, had suffered a truly major defeat. For them, this moment was the beginning of a glorious epoch when the weak ones of the earth, the workers of the world, would have their chains smashed by a victorious Communist Russia.

But Trocmé, Theis, and Darcissac had other thoughts. They too saw Hitler as a monster who had invented and mobilized a great evil; and they too rejoiced at this, his most significant defeat thus far. But for them the killing that had created this great victory over murder and humiliation was itself evil. De Gaulle's Secret Army was an army dedicated to military victory by means of killing; the Communists were an international force eager to use any means, including killing and the hatred that motivates killing, to eliminate institutionalized cruelty from the face of the earth; but though the Chambonnais were friends of France and friends of the weak, the poor people of the earth, for them human life was so precious that they found it impossible to justify the killing that had produced this great victory.

Unlike other groups in the camp, political doctrines were not part of their thinking. An intimate ethical and religious judgment caused their deep ambivalence about the victory at Stalingrad. Trocmé had a desire (as he put it in his notebooks) "not to be separated from Jesus." What this meant to him was that God had shown mankind how precious man was to Him by taking the form of a human being and coming down to help human beings find their deepest happiness. Trocmé believed also that Jesus had demonstrated that love for mankind by dying for us on the cross. And if these beliefs sounded too mysterious, he knew that Jesus had himself refused to do violence to mankind, refused to harm the enemies of his precious existence as a human being. In short, Jesus was for Trocmé the embodied forgiveness of sins, and staying close to Jesus meant always being ready to forgive your enemies instead of torturing and killing them. Trocmé could not bear to separate himself from Jesus by ignoring the precious quality of human life that God had demonstrated in the birth, the life, and the crucifixion of His son.

When, decades later, I asked Édouard Theis whether he and Trocmé believed that the Soviet Union should have used means other than violence to protect herself from the Germans, he answered, "No. They had to use violence then. It was too late for nonviolence. Both the Germans and the Russians were *embarqués*, committed to mass murder—that is, to warfare—and they had to play out their terrible roles upon each other. Besides," he added, "nonviolence involves preparation and organization, methods patiently and unswervingly employed—the Russians knew nothing of all this. Nonviolence must have deep roots and strong branches before it can bear the fruit it bore in Le Chambon. Nonviolence for them would have been suicide; it was too late."

While the cheering was going on, Theis and Trocmé could not express these convictions to their newfound friends. But in the course of the weeks that followed that momentous announcement, the old inmates came to understand that the three Huguenots were brave men who had spent the past three years leading a whole village into stubborn, active resistance against the cruel ones of the earth. Bravery, especially humble, efficacious bravery, not simply inner spiritual fortitude, was for most of the inmates an impressive virtue, and so with every passing day, understanding and warmth increased and made the newcomers the spiritual nucleus of the camp.

4.

Soon gifts started coming to the camp from Le Chambon, brought mainly by Magda Trocmé and by the son of Roger Darcissac. Soon, the shelves the Chambonnais built above their broken-down beds were loaded with packages, so that their area looked more like a grocery store than a prison barracks.

At first the old inmates would stare wide-eyed at each new bit of food or clothing, and they would say, "We didn't know that pastors were so rich!"

To this the Chambonnais would answer, "Oh, we aren't rich. These are from the poor people of Le Chambon, who are used to giving and who love us."

One of the old inmates once said, "I'll be damned! Look, I'm the chief of the Communist cell in Béziers—or used to be—and I haven't gotten a thing from my comrades there. So it's like that in Le Chambon, is it?" And the eyes of the hardened leader lit up in boyish admiration for those Christians who strangely resembled Communists—or at least what Communists *should* be.

"Yes," the preachers went on in their conversation with the man from Béziers, "that's the way it is in Le Chambon. But mind you, not all poor little parishes are like that. Le Chambon is a good parish."

In the course of that conversation, Trocmé thought, I'm being tested here so that I can find out what I really believe and what I really am, but the Chambonnais are being tested too. And they are not being found wanting. All these years I have been their pastor, and I have seen their humble faces, their lowered eyes, and I have heard their hesitant words in the midst of all those things we have done together. But I have never known until now what affection and what devotion have been hiding behind those shy, laconic gestures.

But it was not the gifts that made the Chambonnais the spiritual center of the camp. It was the Protestant services and the discussion meetings. While they were all expecting deportation and death to come at any time, Theis suggested that there be Protestant services in the camp, even though the three Chambonnais were the only three believing Protestants there. The Vichy authorities in the camp agreed to give them a room in the barracks and a blackboard upon which Theis could write in his neat hand the numbers of the hymns to be sung.

At the first religious service there were twelve present, including the Chambonnais. Theis wrote down the numbers on the blackboard, gave a Bible reading, and then Trocmé gave a sermon. After this, Darcissac led them in a hymn, which they sang at the tops of their voices out of the sheer joy of being able to sing aloud together after so long a period of mutterings and fearful solitude. The hymn Darcissac taught them went: "Faith makes the strongest ramparts fall before our eyes; faith lifts the bolts and wins the battles."

After the benediction, the group sat around to talk, and one of the Communists asked, "Your hopes—are they for the next world or for this one? If they're for this world, we're with you. If they're for the next, then they don't interest us—they're too vague, too far away."

To this, Trocmé answered, "Faith works on earth. I do not know about Heaven." For Trocmé, the test of whether a faith was real lay not in patience or in passionately rehearsed imagery but in what that faith could do to make our own lives and the lives of others precious now, in our homes, in our villages. For him, eternity was not as important as efficacy. After an excited discussion about the power of faith to change one's everyday life, the group insisted that they meet not only on Sundays but every evening. They finally agreed to meet three times a week.

On the first evening there were twenty present, and the small room given them by the authorities was full. The next evening meeting attracted forty, and twenty prisoners had to hear and sing while standing outside the open windows of the barracks.

The director of the camp began to get worried: what could they possibly be saying to attract and keep the interest of all these dyed-in-the-wool atheists? And so a policeman became a regular member of the congregation and discussion meeting. He would sit in the front row in a seat reserved for him, and he would take notes constantly, saying nothing.

The main topic of the discussions was the relationship between Christianity and Communism. Since it was dangerous to try to harmonize these two with each other before an officer representing a bitterly anti-Communist government, the group decided to make their discussions sound like a defense of Marshal Philippe Pétain and his National Revolution. They continued to use the name of Jesus—the name was pleasing to the government of France—but instead of using the name of Karl Marx, they used the name of the French chief of state, Marshal Pétain. Now they sounded as if they were reconciling Christianity with Pétain's National Revolution! The officer was plainly impressed by the reeducation the Protestants were bringing about in this camp full of atheistic resisters.

The camp was becoming an organized group of resisters against Vichy right under the eyes of Vichy. After the first week of meetings, many of the lank and grizzled inmates could be seen walking across the area to and from meals or to and from the latrine singing and whistling the melodies of Protestant hymns. Peacefully the camp had gotten completely out of hand, and the representatives of Pétain's National Revolution never knew it. The three leaders were creating another Le Chambon.

5.

One day, after more than a month of incarceration, the three Protestants received a call to report to the office of the director of the camp. On the way to the office outside the barbed-wire fences they thought of the possibility that they might be punished for having let the discussions get daringly critical of Vichy; and they thought that this might be the moment when they would be told that they were going to the death camps of Germany or Poland. But when they arrived, the director told them that they were being released.

They went back to their barracks, packed their things, and distributed to the others the gifts they had left—mainly food, which was much needed in that harsh camp. When they returned to the administrative office of the camp, they were told that they would board the 10:00 A.M. train for Limoges, but that there was one formality they had to observe before separation from the camp: they had to sign a certain paper.

The paper had two parts. One exacted a promise "to respect the person of our leader, Marshal Pétain." The other read: "I shall obey without question orders given me by governmental authorities for the safety of France, and for the good of the National Revolution of Marshal Pétain." Darcissac, who was already working for the government as head of the public school for boys, had already signed such documents; he shrugged his shoulders and signed the oath.

Trocmé, impulsive as he was, was already lowering his pen to sign the paper when Theis pointed out the second part of the oath, about obeying the orders of Pétain without question. The first part, about respect for the person of Pétain, was something both ministers could endorse without hesitation; they respected the person of every human being. But when Theis with his less impulsive approach put his finger on that second part, Trocmé said, always speaking for both of them, "We cannot sign this oath. It is contrary to our conscience."

During this discussion the director of the camp had been in an adjoining office, but now he stood behind them, shouting, "What is this? This oath has nothing in it contrary to your conscience! The marshal wishes only the good of France!"

Trocmé answered, "On at least one point we disagree with the marshal and his National Revolution: he delivers the Jews to the Germans and thus to death. We are opposed to such action. When we get home, we shall certainly continue to be opposed, and we shall certainly continue to disobey orders from the government. How could we sign this now?"

The director sputtered with anger. "You refuse to sign! This is insane! You know as well as I do the vicious activities of the Jews. They're the rottenness in France that we must cut out of our body politic."

Then he forced himself to subdue his patriotic zeal and to forsake the rhetoric of the National Revolution. He suddenly became confiding. "Look, be reasonable. I appreciate your courage, but this is— Look, you have wives and children. Sign. It's just a formality. Later, no one will notice what you did here."

Trocmé replied, "If we sign, we must keep our word; we must surrender our consciences to the marshal. No, we will not bind ourselves to obey immoral orders."

"So be it," the director said. "You'll rot here indefinitely, if the

Germans don't deport you first. Take these men back to their barracks!"

Darcissac had signed because he had already endorsed such an oath as head of the public Boys' School in Le Chambon, and also because he did not want to lose his job. But when he heard that the ministers were not leaving because of the oath, he begged to be kept in the camp with them. But the director would not let him remain. And so he said good-bye to his friends, who told him that they understood his position and that his was not a calling like theirs.

Darcissac had relatives and friends in Limoges, and decided to spend some time with them on the way to Le Chambon. When he got off the train in that great city, he had to cross a wide park to get to his uncle's home. The park was empty that day, except for a solitary mule whose master had left him behind to graze. With his fears for Trocmé and Theis strong in his mind, Darcissac almost ran up to that mule and embraced him, screaming, "I'm free! I'm free!"

After Darcissac left, the two ministers were processed again—down to the measuring of their noses. When they reentered the two ranks of barbed wire, Trocmé knew a moment of fear; he felt that he and Theis had now committed suicide.

6.

Trocmé had touched upon an important point when he expressed his disagreement with the National Revolution in terms of delivering foreign Jews to the Germans. This delivery—and anti-Semitism—were important to Pétain's plan to placate the Germans and to strengthen the "soul" of France. In the fall of 1940, only a short while after the defeat of France, Vichy, on its own initiative, but with an eye on its conquerors, took a census of foreign Jews in France. This immediately made the capture and deportation of Jews a neat bureaucratic procedure: the government had names and addresses, and the Jews were in their hands. In the first few days of October 1940, Jews were excluded from elected bodies, from the higher ranks of the civil service, from the courts and the military, and from all positions that could wield a strong public influence.

Even before October, in August 1940, a law that penalized anti-Semitic diatribes in the press was repealed. And restrictions on the Jews became more and more severe as the years of the Occupation went on. For instance, by the summer of 1942, about twenty thousand Jews had been incarcerated in French concentration camps in the Southern, or Unoccupied, Zone of France alone, imprisoned only for the crime of being Jews. Vichy did not intend to *destroy* the Jews, as the Nazis intended to do, only to eliminate them from the body politic of France, and Pétain himself refused to let the Germans require that Jews in the Southern Zone wear the yellow star that identified them for hatred, ridicule, and capture in the streets. But Vichy singled them out for contempt, deprivation, and imprisonment, and thereby helped the Nazis greatly in their program to destroy the Jews of Europe.

The National Revolution had as one of its overarching ideals a smoothly running machine purified of all recalcitrant grains of sand, free of all "dangerous anarchists," as the director of the camp had put it. And among these forces for anarchy that would disturb the smoothly oiled functioning of this machine were the Jews. The unlimited authority of Pétain and the seamless unity of France demanded, among other things, the sequestration of the Jews, economically, politically, culturally, and physically in the brutalizing concentration camps that extended across the body of France.

As Cardinal Gerlier, archbishop of Lyons and chief primate of France, put it in a speech of September 6, 1942, the National Revolution of the early 1940s had brought to France a "new order . . . built on violence and hatred." The Jews were not the only victims of this new order—Communists and Freemasons were among them—but it was the Jews who came to Le Chambon across the great mountains of southern France and who impressed their pain upon the minds of the Chambonnais.

All of this—and Trocmé's belief in the priceless preciousness of all human life—was what lay behind—no, pervaded—Trocmé's statement to the director of the camp at Saint-Paul d'Eyjeaux: "On at least one point we disagree with the marshal and his National Revolution: he delivers the Jews to the Germans and thus to death. We are opposed to such action."

7·

When they returned to the barracks, their friends, thinking they had come back to bid them good-bye, greeted them with joy: "You're free. You're free! Bravo!" There was not a shadow of envy in the eyes of these men who had been imprisoned for so long and who were going to continue to live under the moment-to-moment threat of deportation.

When the ministers told them that they had refused to sign an oath of allegiance to Pétain and so were not being released, the old differences among the inmates reappeared. A Communist said, "What? Refuse to sign a scrap of paper that will free you to resist the Fascists and defend the poor? This is weakness. This is idiocy." An old railroad man gently told them as he helped them rearrange their beds, "If it had been me, I would have signed with both hands. And then the day after tomorrow, *up* would go their railroad tracks—*boom!* With both hands I would have dynamited them to Hell."

Another inmate said, "Now you must really see that your Christian *truc* [dodge] doesn't work. A revolution by nonviolence and nonlying—that might work with decent people, but with skunks? Here's what will happen. You say, 'I'll not sign your paper,' and they'll keep you here till the Germans come to get you. To succeed, you've got to be a skunk with the skunks. You can't survive otherwise, let alone *do* something!"

The ministers were quiet. They were still depressed from their farewell with Darcissac, who had left them the way a healthy person might say farewell to a friend dying in a hospital room. Trocmé thought, It's true. I've condemned myself and Theis to death.

The next morning, they were called back to the administration hut and greeted by the director himself. "I have good news for you," he said. "I have just had a telephone call from the office of Pierre Laval. I have been told to free you immediately." (Laval was the controversial, powerful second-in-command to Pétain at Vichy.)

"But we cannot sign the oath of loyalty!"

"Never mind," said the director. "I have orders from the top to free you two without your signatures. You must have some good friends way up there. Anyway, get ready. The train leaves at ten o'clock this morning, and I don't want any more trouble in this camp. Pack your bags."

The release of the ministers is a mystery; to this day, nobody in Le Chambon knows with certainty why the pastors were set free. Marc Boegner, head of the Reformed Church of France at that time, has since claimed that he spoke to Marshal Pétain on the ministers' behalf. It seems that the BBC had been making propaganda about how harshly Pétain was treating Protestant ministers and about the comparison between Pétain's persecution of the Protestants and Hitler's attacks upon the German Protestant churches. The battle of Stalingrad had been won by the Allies, and it is possible that Laval and Pétain wished to please the Allies in the event that the Axis lost, just as they had been trying to please the Nazis when they believed that the Germans would conquer Europe and the world. Also, the director of the camp might have told Vichy of his fear that the Protestants were beginning to organize the camp in mysterious and possibly dangerous ways. It might have been a combination of some or all of these causes, or it might have been none of them, but as the two big men approached the barracks for the last time, they were not concerned with this problem. They were at peace with themselves.

When they told their friends that they were really free, and without signing the oath, the one who had recommended being a skunk with skunks said, "Shit! Your Christian dodge works pretty well sometimes, doesn't it?"

Before they left, the ministers gathered together their closest friends, formed a circle holding hands, and sang a song Roger Darcissac had taught them. The song was one the people of Le Chambon often sang. Its melody was the same as that of "Auld Lang Syne," and its words meant: "It's only *au revoir,* my brothers; it's only *au revoir.*"

The paper pledging allegiance to Pétain and the National Revolution was never again used. There had been a little breakthrough at Saint-Paul d'Eyjeaux.

But none of the circle, except for the ministers, lived to go home. A few days after the Chambonnais left, all of the remaining inmates were deported to concentration camps in Poland and salt mines in Silesia. Almost all of them died at hard labor or in the gas chambers of a death camp.

2

André Trocmé, the Soul of Le Chambon

1.

Vichy and the Germans were quite right to arrest Trocmé first in their attempt to clean out that "nest of Jews in Protestant country." It was Trocmé more than any other single person in the village who had made what happened happen.

A refugee who spent many months in Le Chambon once found herself reminiscing about her arrival there. "Upon my arrival," she said, "I saw Pastor André Trocmé, and I knew instantly that he was the soul of Le Chambon."

There are differences of opinion among Chambonnais about various things that happened in those four years, about where the blank ration and identity cards came from, the cards that saved so many refugees' lives, and about whether there were Chambonnais informers sending information to Vichy about the Resistance in the first two years of the Occupation, and about other topics. But if you walk into any house in Le Chambon that still contains a person who was an adult during the war years, and if you ask that person, "Why did Le Chambon do these things against the government and for the refugees while the nearby Protestant village of Le Mazet did not?" you will receive only one answer: "It was Pastor André Trocmé." Whether or not this can be proved to be true (and who can answer such questions with finality?), the fact is that when you try to understand the peculiar spirit of Le Chambon, you find that all roads lead to André Trocmé, just as all roads led to him when Vichy was trying to punish and wipe out the Resistance in Le Chambon in February 1943.

The refugee who saw Trocmé as the soul of Le Chambon meant many things by her metaphor, but mainly she meant that he was the quickening spirit, the warming force in that gray little village during those poverty-stricken and dangerous years. She meant that he set an example for the Chambonnais: he "welcomed refugees in a wide embrace," as she put it. A soul for her was a source of outgoing warmth. A Chambonnais, not a refugee, and one of Trocmé's oldest friends, once said, "He wanted to hug you—yes, even to kiss you. My! You know, he was—yes, he was sexual, almost, he wanted so to embrace you!" True, the speaker was himself a very shy man, and his feelings about Trocmé's almost erotic warmth reflect as much about him as they do about Trocmé, but the pastor's open-armed generosity toward his fellowman was as real as this man's shyness.

Trocmé was far from being only a loving man. Everybody who knew him at all well agreed that he was capable of immense anger. In any case, nobody I have ever talked with has suggested that Theis, the assistant pastor of the village, be thought of as the soul of Le Chambon, and the reason why this is so can be seen in two stories, stories that reveal the immense differences between a man of enlivening and sometimes terrifying passion and a man who was a rock upon whom others could build lasting edifices.

In the fall of 1938, Édouard Theis was beginning his life in Le Chambon as half-time pastor under Trocmé. One evening he was asked to officiate at his first funeral services in the big, gray, boxlike temple. Coming in through the main door with the rest —a door over which stands the inscription "Love one another" —he quietly sat down among the parishioners and waited. After a while somebody said, "Well, where is the pastor?" Then, and only then, Theis lumbered up to the front of the congregation and in his quiet voice said, as if he himself were still in some doubt about it, "Oh, but that's me. Yes. I am the pastor."

The other story is about André Trocmé. Long after the war, he was giving one of his many lectures on nonviolence outside of Le Chambon when a certain member of the audience began to turn to his neighbor and whisper audibly whenever Trocmé made an interesting point. It turned out later that he admired what Trocmé was saying. After a few such whisperings, Trocmé stopped his lecture suddenly, walked up to the vivacious whisperer, fixed his blue eyes down upon his upturned, stupefied face, and shouted, his fair, baby-soft face flushed with rage, "Out of the room! Get out of the room!" And with a massive arm he pointed the way to the door while calling over his shoulder for an usher to conduct the miscreant out. The lecture had been on nonviolence.

Magda Trocmé once called her husband a turbulent stream, thrusting its way with great speed and force through and around obstacles, changing always as it struck and flowed. Then she pointed to a rock in such a stream and said, "Ah, that is Édouard Theis. The rock of Le Chambon. Look. It sits for hours and days and years, faithful to where it stands, unshakable." A rock is not a source of movement. A stream is; in fact, it is itself liquid turbulence. And a soul, whatever else it may be, is a source of vivacity. It brings passion to a body and moves other bodies and souls with that passion. Such was André Trocmé.

But a soul is not only emotion. For the Greeks it was an *arche* (initiator, source). The rocklike resister of Vichy and the Nazis who spent the last year of the Occupation taking Jewish refugees through dangerous mountains to safety in Switzerland was not a bold initiator. Theis once said, "I was the follower, the helper, yes! The second fiddle to André Trocmé!" And his happiness in saying this was not only the happiness this man feels in anonymity, but the happiness he feels in telling a plain truth.

For the early Greek philosophers, the soul was the initiator of the body's movements, the one world-moving force, which leaves the body in death, but in life initiates, say, each movement of your eyes across this page, each movement of your hand to turn it. The nearby village of Le Mazet was "dead" as far as attracting and protecting refugees was concerned because it lacked—perhaps among other things—the unflagging inventiveness of the mind of André Trocmé. He kept discovering new things to do that would give substance to the words above the temple door: "Love one another."

Since ethics is concerned with an individual's character, it is fitting that our ethical attitude toward Le Chambon be directed, ultimately, toward an individual. There was no more creative individual in Le Chambon than Trocmé. It is true that he could not have done what he did without Theis, and his wife, and many others, just as a soul cannot do what it does without a body, at least here on earth; but it was he who set the goals of Le Chambon during those years, and it was his practical ingenuity that set the example for successfully pursuing those goals.

Instead of trying to pluck out the heart of his mystery, *the* key to his mind, let me try to answer a rather modest question: What events in his life can help us to understand his belief that human life is priceless? It was in the service of this belief that he made Le Chambon a "city of refuge" (a term taken from the Old Testament), and if we can understand that belief in terms of his life, we shall be ready to understand in depth what he did in Le Chambon.

André Pascal Trocmé was born on Easter of 1901 (his middle

name is the French word for Easter). When he was born, the big windows of the bedroom were thrown open so that the sunlight of Easter morning could come through, and so that the bells of the nearby basilica celebrating the resurrection of Jesus Christ could bring their sounds deep into the room. (Those bells were later melted down by the Germans to make cannons.) From the beginning, his pious Protestant parents hoped he would become a pastor.

The city of his birth was Saint-Quentin, in Picardy, the northeastern region of France bordering on the English Channel and Belgium. In this region John Calvin, whose French Protestant followers came to be called Huguenots, was born in the sixteenth century. Here some of the most violent and protracted battles of World War I would be fought. Religion and war would be the main forces in Trocmé's youth.

His mother, the former Paula Schwerdtmann, was born in Germany of German parents and ancestry. His father was Paul Eugène Trocmé, a descendant of an old Huguenot family. In his early youth, Trocmé was to find his German relatives far warmer and more attractive than the Huguenot kinfolk of his severe French father. In fact, far from leading the provincial life typical of the tiny minority of Protestants in twentieth-century France, Trocmé led a life that was international to its very roots. Sometimes that internationalism took the form of having his German grandfather put his gigantic, warm hands upon his head and bless him in order to keep away the *Bösewicht* (Devil). But sometimes his internationalism caused him great pain. During World War I, he saw the countrymen of his mother kicking and shooting Russian prisoners in the streets of Saint-Quentin.

Trocmé's father was a very successful manufacturer of lace, whose success in business had contributed much to the wealth of Saint-Quentin. Their house had a dozen bedrooms, and because of his father's unexplained orders, its second floor was to remain remote for him until he was thirteen years old. As a child he played in a large yard surrounded by high stone walls. His early years of schooling were spent at home with private tutors, and one of the most striking impressions of his youth was that he was one of *le peuple Trocmé* (the Trocmé people), who were apart and different from others. He was separated from others by wealth and by the rigid demands his father made upon his behavior and thinking. In his childhood his warmhearted, expansive German relatives only contributed to his feeling of separation from his native city. When he wearied of playing in the walled yard with his carefully chosen friends or his brothers and sisters, instead of leaving the yard, he stayed within those walls and yearned for his vacations with his pipe-smoking, beer-drinking, unbuttoned German relatives. But even between him and them there was a wall; his father did not approve of smoking, drinking, or bending, and so André found himself loving his German relatives but not entirely approving of them. They were not in that elite group of "Trocmé people."

Before World War I, two events smashed holes in the walls that separated *le peuple Trocmé* from the rest of mankind. In the fall of 1912, André was playing at war with his nephew Étienne; pebbles were the cannonballs, cardboard tubes were the cannons, and the *boom* came from the boys themselves. There was a back door in the walls of the Trocmé estate that the gardener used to enter and leave without disturbing the family. This door Étienne and André romantically labeled the "postern gate," their own secret entrance to their own medieval fortress. On this fall day, the boys were so busy playing at war that they did not notice that the postern gate was slowly swinging on its hinges—the gardener had forgotten to lock it. But at last the swinging gate attracted their attention, and, stopping their play, they saw a bony, pale man wearing a flat cap, a short coat, and shapeless gray pants. A cigarette hung from his lips. He looked at the two upper-class boys for what seemed to be a long time, in silence. Then he started shaking his head, and a glance of bitter but detached pity came into his eyes as he said, *"Tas de cons"* ("Bastards"). Then he left, closing the postern gate behind him.

For the rest of his life Trocmé remembered that *pâle voyou* (pale guy), and for the rest of his life he would have to bear the burden of those looks and that judgment. From then on, he knew that *others,* not the "Trocmé people" but the poor, the excluded, the bitter ones of the earth were watching him and judging him. The wall had made his life a lie because it had hidden him from the "pale guys" and had hidden the "pale guys" from him. As much as any other single event in his life, this incident caused Trocmé to turn his back upon his class and to work with the poor.

The second major event before World War I was of even deeper significance to him than the look of the "pale guy." On a beautifully flowery Sunday in June 1911, André Trocmé's father killed his mother—or, at least, became morally responsible for her violent death.

Two months after his tenth birthday, Trocmé joined the family on an auto trip to their country home. Two of his brothers and a first cousin sat with him in the back seat, while his mother, wearing a veil, sat with his father in the front. His father had chosen to be chauffeur for the day and had decided to take the long road to their country place so that the family could see the fields in flower. At the very beginning of the trip, at the edge of Saint-Quentin, they found themselves at a closed level railroad crossing, and Papa started to get angry. As he slowed down approaching the crossing, his anger suddenly mounted—a ramshackle little car slipped around and in front of him. When the crossing was clear, the little car, as if out of spite, spun its wheels and threw clouds of dust upon the shiny limousine of the Trocmés. Paul Trocmé was furious. They came to a descent, and the *tacot* (jalopy) started to slow down, holding the exact middle of the road. The children in the back seat begged their father to sound his horn to get the little car to move over to the right. His frustration became uncontrollable when he realized that his horn was not working. He was going to pass that car.

As he came around it to the left, his wife took his arm and called out, "Paul, Paul, not so fast! There's going to be an accident!"

Suddenly there were the screams of people in terror, the feeling of a hammer blow on André's head, and then nothing but the crickets sounding in the fields and gasoline falling drop by drop somewhere nearby. The family pushed themselves out of the twisted iron of the car and stood in the road, with Papa holding his broken right fist in his left hand. They were all trembling and laughing wildly at being alive.

Then one of them saw Mama. But it was no longer she. It was a body on the road about thirty feet from the car, to the rear, a large body covered with dust, lying on its back with its legs slightly apart and with a thread of blood trickling down from the right corner of its mouth. The eyes were closed, and on the face, which was pointing up to the sky, was a proud, indifferent look, the look of what Trocmé, in retrospect, called the "thing," death.

Three days later, she was declared legally dead, but she never recovered consciousness, and for her son she had died on the road, sprawled on a little hill. When his father announced to the children, "Your mother is no more," André hugged him, and in his new ascendancy over his authoritarian father, he said, "Daddy, promise me you'll never have another car."

His father screamed, "I killed her! I killed her!" And Robert, André's older brother, came to him, embraced him, and kissed him.

Until the moment when he saw his mother sprawled on the rise in the road, his home had been "deep as a cradle," to use his own phrase. A heavy-legged, clumsy, big blond boy, he had felt especially close to his mother. Later, when he remembered his relationship with her, it seemed that he had not existed in his own right before she died. He was a part of her soft body, swept into her being by her kisses, by the music she played for him on the piano, and by the love she bore him. Now he was alive and she was a "thing." Death and solitude became for the first time in his life—and for the rest of his life—as real as anything else in the world, as present, as insistent, and as close to him as his need to take the next breath of air.

Trocmé never knew whether there was life after death, whether there was a Heaven or a Hell where souls separated from their bodies go to spend eternity. Later his honesty about his ignorance would hamper his ministry, since he would not console survivors with assurances about meeting their loved ones "up there." Death for him was loss, the loss of a dear person now replaced by the "thing."

Here is how he summarized, many years later, the effect of her death upon his life:

> If I have sinned so much, if I have been, since then, so solitary, if my soul has taken such a swirling and solitary movement, if I have doubted everything, if I have been a fatalist, and have been a pessimistic child who awaits death every day, and who almost seeks it out, if I have opened myself slowly and late to happiness, and if I am still a sombre man, incapable of laughing whole-heartedly, it is because you left me that June 24th upon that road.
>
> But if I have believed in eternal realities, in things that are beyond The Thing, if I have thrust myself toward them, it is also because I was alone, because you were no longer there to be my God, to fill my heart with your abundant and dominating life.

What an expression this is of a man's awareness that a human life is infinitely precious! Just as suffocating can show us the preciousness of air, her death showed him the preciousness of her life.

André Trocmé firmly believed that his father was morally responsible for her death: in effect, he had killed her, as far as the son was concerned. But the son still loved the father, and he hugged him and forgave him when he cried, "I killed her! I killed her!" Trocmé's first encounter with death was at the same time an encounter with his need to forgive lovingly the "killer." In the same event he learned the preciousness of the victim's life and the preciousness of the slayer's life. For the rest of his life—except for one moment in 1939, when he thought of assassinating Hitler—he would eschew the vicious circle of revenge. The loss death inflicted was too awesome to be perpetrated upon any human being. Life was too precious—all human life.

2.

André Trocmé grew to young manhood within the barbed-wire boundaries of a city about twenty miles from where the battle of the Somme took place in July 1916, when he was fifteen years old. Between September 1914 and February 1917, the German Army, which was occupying the city of Saint-Quentin, almost bled the city to death. Inside the barbed wire that surrounded the city stood German sentinels who demanded signed permits from anyone wishing to leave. And there were many who wished to do so; the food supply was utterly inadequate, and people were trying to leave the city in order to buy food from the surrounding farms.

In February of 1917, the Trocmé family left Saint-Quentin as refugees from an almost dead city. They entered Belgium, and André learned the pain of hunger and the dark misery of begging food from the poorest of the poor. But he learned also what integrity was. For a while he was left to study the sciences with a Belgian Catholic priest. Being interested in religion, he tried to talk about that subject with the priest. Each time he tried, the priest became red with anger and immediately left the room. He had made a vow that he would not talk about religion with his young Protestant charges, and he was going to keep that vow.

Years later in Le Chambon, Trocmé would have Jewish children studying with him and with other Protestants. In the spirit of this priest, he would encourage the Jewish children to observe their own holidays, and he—as well as Theis—would refuse to allow any children to be converted to Christianity behind the backs of their parents. His relationship with the priest was the beginning of a long train of experiences that would result in his adopting the following principle: Help must never be given for the sake of propaganda; help must be given only for the benefit of the people being helped, not for the benefit of some church or other organization that was doing the helping. The life and the integrity of the person helped were more precious than any organization. And so Trocmé would never try to convert the Jewish refugees who came in need to Le Chambon.

One of the reasons for his curiosity about religious matters was that in his father's house there was, in the end, only one prayer: "Teach us to do our duty." Morning, afternoon, and evening there was always prayer in the mansion, and there was frequent Bible reading and meditation. On Sundays the older brothers would go to the temple twice. But theirs was a religion of duty toward a distant God. "Teach us to do our duty" became a formula that kept the individuals in the household from communicating their own feelings to each other, or seemed to make it unnecessary for them to do so. From the lips of that severe Huguenot, his father, the prayer came to the son like a word heard from a great distance in cold, crisp air across acres of snow. The sound was clear, but the feeling was not there.

In the midst of the war, and in the desert of his own solitude, André Trocmé had become a member of the Union of Saint-Quentin. The union was a Protestant organization of young people, almost all of whom were the children of laborers. They met in a bare old room above the entrance door of the temple; all they had was a table and chairs. They conducted their own services in the intimacy of their friendship, either at the table talking about the Bible, or on their knees, often in tears, praying aloud to God to be saved from lying or from sexual impurity. When one of them thanked God for deliverance from some sin or some misfortune, the rest of the group gathered around him in a shared joy. If a problem persisted, they begged God to increase their love for each other so that they could help the miserable one to recover his spiritual well-being.

For André Trocmé, this place and this group were paradise on earth. He no longer had to hide his deep fears or regrets from stern, powerful adults. He could tell everything to his young friends and to their God.

Looking back as a sixty-year-old man to those months in the union before the Trocmés became refugees from Saint-Quentin, he was sure that he believed more in the union than he did in the power of God. Here he learned to feel the power of human solidarity. It was to the union that he preached his first short sermons with his elbows squeezed tightly against his big-boned, clumsy body. And it was in the union that he conquered day by day the timidity that solitude had created in him. Around him now was not an alien world but young people in passionate communion with each other's fears and hopes.

But the adolescents in the union did not erect their own walls. In the autumn of 1916, after the Germans had held their positions in the battle of the Somme nearby, the occupants of Saint-Quentin started to build the Hindenburg Line, the precursor of the French Maginot Line of World War II. In order to do this, they used Russian prisoners, who lived in cold, starvation, and filth in camps situated inside Saint-Quentin. Hungry themselves, the Germans starved their prisoners cruelly while working them to death to build their line of subterranean fortifications. It was not an extraordinary event when one of those Russians, thin as skin on bones, dropped in a faint upon the cobblestones and was either kicked by a German soldier or shot through the head on the spot.

It was forbidden to do anything to help them, but the teenagers of the union did so nonetheless. Every day they brought a large potful of vegetables they had managed to collect into the prison camp (with the help of a compassionate German sentinel), and they distributed the food to the Russian prisoners. Before they left, they passed among the prisoners with cigar butts and fistfuls of tobacco they had found in their homes or on the streets or the floors of public places. Occasionally one of them was caught and imprisoned for eight days for having communicated with the Russians. Always they left the camp hating the Germans. Even the half-German André, who had loved his own German relatives so gratefully for *their* love, could not suppress his angry hatred against them.

He had not yet found a consistent attitude toward all human beings, had not come to regard all human beings as precious, but he had departed forever from the "Trocmé people." For the rest of his life he sought another union, another intimate community of people praying together and finding in their love for each other and for God the passion and the will to extinguish indifference and solitude. From the union he learned that only in such an intimate community, in a home or in a village, could the Protestant idea of a "priesthood of all believers" work. Only in intimacy could people save each other.

And because he learned this well, the struggle of Le Chambon against evil would be a kitchen struggle, a battle between a community of intimates and a vast, surrounding world of violence, betrayal, and indifference. Le Chambon would be, at least during the first four years of the 1940s, the union he sought.

3.

During the war, before the Trocmés were to leave their city as exiles, Saint-Quentin was like a sprawling hospital. The smells of chemicals and of rotting flesh were everywhere, and at night, trains full of bodies from the front crossed the city to the places where the bodies were to be incinerated. With all this, the hatred of the French toward their German occupiers grew more and more bitter; but in the midst of all this, André Trocmé began to have a fundamental, single attitude toward mankind—including the enemies of France, his mother's compatriots, the Germans.

One day he saw coming toward him a straggling column of wounded German soldiers. The Germans were losing the war, and, lacking transportation, the wounded had to drag their broken bodies step by step to the hospitals assigned to them. In the first row of the column, the seventeen-year-old boy saw three heavily bandaged men. The man in the middle had, instead of a head, an enormous ball of bandages. He probably could not see, because he stumbled and was being led by his comrades. When he came closer, the boy saw that he had no lower jaw. In its place there was a ball of linen, and from this ball there hung clots of blood.

André Trocmé had played at war in the walled garden, and he had heard war being discussed as if it were an heroic duel of honor between good and evil, a duel in which the honor, courage, and skill of one adversary sent the other down to deserved defeat. Now that the Germans were losing, this idea of war was a mania in Saint-Quentin; around him there was not only hatred for the crumbling enemy, but triumphant contempt. But he could not hate nor could he despise that man without a jaw. And for the first time in his life he found his hatred turning not against the enemy but against the war that had wounded that particular man so terribly. All he could think of as he looked at the blinded, stumbling monster was, Look there, see what you have done to your brother.

A few days later, he met a German soldier on the staircase of his own house, part of which was being used as military quarters. The German stopped, looked kindly at the lad, and touched his arm. "Are you hungry?" he asked in German, and offered him a bit of *Kartoffelbrot,* the black potato bread of the German Army.

"No," André answered in German, "but even if I were hungry, I would not take bread from you because you are an enemy."

"No! No! I am not your enemy," the soldier said.

"Yes, you are," the young man persisted. "You are my enemy. You wear that uniform, and tomorrow you will perhaps kill my brother, who is a French soldier fighting against you, trying to get you Germans out of our country. Why have you come into our country carrying war and suffering and misery?"

"I am not what you think," he answered. "I am a Christian. Do you believe in God?"

The boy's face brightened slightly—the man was using words he had often heard and uttered throughout his young life.

"At Breslau we found Christ," the soldier went on, "and we have given him our life." Then he told Trocmé about a certain sect to which he now belonged.

The soldier said, "Men cannot hurt those who have put all their confidence in God. One day a man who hated the work of our sect came into the meeting hall to kill our leader, but his pistol misfired, and we all knew this was a sign from Heaven."

Standing there on the staircase with his hand on the young man's arm, he went on, "I shall not kill your brother; I shall kill no Frenchman. God has revealed to us that a Christian must not kill, ever. We never carry arms."

"But how can that be?" the boy asked. "After all, you are a soldier."

"Well, I explained all this to my captain, and he has allowed me to go into battle without arms. Usually, telegraphers like me carry a pistol—or a bayonet, at least. I have nothing. I am often in danger when I am in the lines, but then I sing a hymn and I pray to God. If He has decided to keep me alive, He will. If not . . ."

André Trocmé had met his first conscientious objector. Perhaps if the soldier had been French, the boy would have been indignant at him for refusing to defend his country when André's brother was out there fighting for it and for his own life. But here was a German simply refusing to do what he saw as an immoral job. The courage and faith of the man were plain, and the boy invited Kindler (that was his name) to come to the union for the next Sunday service. Kindler accepted the invitation.

Earlier in the war, the boy had walked across Saint-Quentin with some of his German relatives, shamefacedly speaking German with them before his French compatriots. His warmhearted relatives had come to the city toward the beginning of the German occupation of Saint-Quentin to bring the Trocmés much-needed food and clothing. These walks had been an agony for the patriotic young Frenchman. Now he was walking across the city at the side of a uniformed German soldier. But something was different—he was beginning to feel that every human being embodied something precious.

When they entered the bare hall of the union, his companions showed their surprise at seeing him bring a German soldier to their services. But when he explained, in the simple language of Kindler himself, that this man was a true Christian, and that he would kill no one because he obeyed Jesus Christ, they immediately adopted Kindler as one of their number, like the believing children they all were.

After the service, Kindler spoke to them briefly about his conversion in Breslau and about his life in the front lines. André Trocmé translated. Then Kindler taught them a little hymn that Trocmé never forgot. The hymn was not hard to remember. It went:

> Hallelujah Hallelujah,
> Hallelujah Amen.
> Hallelujah Hallelujah,
> Hallelujah Amen!

Perhaps this was the hymn Kindler sang in battle. In any case, in a moment they were all singing it together at the tops of their voices, like eager, happy children.

Then they all knelt down together on the bare floor and prayed (Kindler in German). This was the first time in André Trocmé's life that he told his most intimate thoughts to God in a loud, clear voice. The German's love and courage had kindled in him a love and a courage that had been waiting for a spark to ignite them.

After the simple Protestant ceremony, Kindler gave him some papers and other private possessions and said that he had to go to the front but he would try to return to pick up his things. "If I am wounded," he said, "or if I am made a prisoner, you will hear from me. If I return home, you will hear from me, too. But if you do not hear from me, send these things to my wife at the address I have written on this paper. If you do not hear from me, it is because God has judged it right to take me unto Himself."

No word ever came from Kindler. After a while, the lad sent Kindler's possessions to his family.

The attitude of nonviolence toward all human beings came to André Trocmé from many sources: his mother's death, which showed him the horrible power of death, his friendships in the union, the sight of that poor monster of a German with a jaw of rags from which hung clots of blood, his own reading of the Sermon on the Mount, and many other experiences. But in its depths his nonviolence stayed as simple as Kindler's; it was an attitude toward people, not a carefully argued theological position. In its depths it was personal; it had to do with the persons he had known, and these persons were mainly his mother, that stumbling monster, and Kindler. Years later, he would study theology in Paris and New York, and he would work to develop persuasive arguments for pacifism. But this work would be primarily for the sake of convincing others. In his own mind, nonviolence was completely expressed in words as simple and direct as Kindler's when he said to the boy, "One must refuse to shoot. Christ taught us to love our enemies. That is His good news, that we should help, not hurt each other, and anything you add to this comes from the Devil!"

4.

After the war, with Saint-Quentin in ruins, the Trocmé family moved to Paris, where André passed his *baccalauréat,* the examinations that qualified him for work in the universities of France. He studied theology at the University of Paris (and there met Édouard Theis), joined an international pacifist organization, the Fellowship of Reconciliation, and began his work with unions in the suburbs of Paris. As in Saint-Quentin, the unions around Paris were usually groups of poor people who prayed together and worked together. Always, because of the "pale guy," he stayed clear of the wealthy bourgeoisie among whom most of his relatives moved.

Having won a scholarship to Union Theological Seminary in New York City, he decided to go to America to study the practical, optimistic Social Gospel that was then dominant there. His decision to go was influenced by a desire to live a life apart from the "Trocmé people."

But once in New York City he found the Social Gospel too secular, too rational for his deeply devout mind. The Social Gospel may be summarized by George Bernard Shaw's quip: "The only trouble with Christianity is that it has never yet been tried." It was an attempt to turn people away from useless words about the "nature of God," and about some distant "Heaven," and to make them give their attention to trying to make this world, this life, a scene of loving, rewarding cooperation among human beings. It wanted men and women to use the physical and social sciences to make human beings masters and possessors of nature, and conquerors of disease and poverty. And it despised those religious leaders who were content to leave society as corrupt as they found it, if only they could be permitted to conduct their services and their theological ramblings in the odor of sanctity. As much as the Marxists did, they despised the religion that is the opiate of the people, that makes people passive instruments and victims of a society that exploits the poor.

On all these counts, as he would agree with the Marxists in the internment camp near Limoges, so Trocmé agreed with the proclaimers of the Social Gospel. For him, religion was a revolutionary force, driving people to bring loving cooperation into every aspect of social life. But both the Communists and the Social Gospel people lacked one element that was central in André Trocmé's mind: the person-to-God piety that every adolescent in the Union of Saint-Quentin felt, and that Kindler felt. For Trocmé, only this intimate relationship between a faithful person and God, only a person's conscious obedience to the demands of God, could arouse and direct the powers that could make the world better than it is. All their talk about the power of the sciences to transform the world into a paradise was empty talk for André Trocmé. For him there was no paradise where God was not to be heard "walking in the garden in the cool of the day."

And so André Trocmé was alienated and lonely at Union Theological Seminary in 1925. Besides, he found the American language a *bouillie* (soup) of melted-together sounds emerging from somewhere between the throats and the noses of these all-too-optimistic and all-too-worldly people. He gave up trying to distinguish their way of pronouncing "Newark" from their way of pronouncing "New York," and he took a job as a French tutor to the children of John D. Rockefeller, Jr. He took the job not only because he needed money—he was financially independent of his father—but also because he thought teaching American boys French might help him with his own American English. And sure enough, Winthrop and David Rockefeller, the boys he was tutoring, helped him a great deal toward understanding and speaking the language.

His brief relationship with John D. Rockefeller, Sr., frustrated him in the special way that the Americans at Union Theological Seminary were frustrating him with their Social Gospel. He was touched when on Christmas of 1925, after dinner, the founder of Standard Oil sang with a shaking voice and tears running down his cheeks some old Baptist hymns (one of them was "He Leadeth Me"). The singing seemed to open up the wells of a deep piety that lay in the man, a piety for which Trocmé was thirsty even after only a few months in America.

But to talk with "Grandpa" was a different matter. Trocmé remembered him as an almost fleshless body seated in a red chair and wearing a white wig. The first thing he ever said to the young man was: "You Frenchmen don't pay your debts. I always pay off my debts." (France had not yet paid off her war debts to the United States.) Then the old man told him that if he would like to be rich, he must always pay his debts. Taking a dime from his purse, he told Trocmé that he was now establishing his fortune.

"Promise me," he said, "that you'll invest this money as soon as you are back home."

At the end of this first meeting, the old man said to him, "More gasoline and higher prices—*that's* what we want."

What a strange doctrine he has! thought Trocmé. He never could understand those Americans—religious or not—who believed that when they had succeeded, they had learned how to solve all the problems of the world. The advocates of the Social Gospel were an instance of this strange belief.

But not all of his encounters with the social ethic of success were this baffling. It was in New York City in the fall of 1925 that he met Magda Grilli. When he first met her in the cafeteria of International House, he did not find her beautiful, but he did notice her candor and her simple-hearted, intelligent way of addressing people. But as the days passed and he began to see her as magnificent with her bright forehead and her deep, dark eyes, he began to fear her presence at the same time that he was yearning to see her.

At the age of twenty-four he had had no sexual experience, and he was committed to remaining chaste until marriage. Chastity was vital to his religious life, not only because he was imitating Christ's chastity but also because chastity tested his Trocméan self-control. He had a boiling, sensual nature and often used showers, baths, and exercise to help him keep it from spilling over into sexual action. Once, in Paris, he had thrown a bucket of dirty water on the head of a girl who was pursuing him. Usually he dared not even look at women, especially attractive ones, except in his dreams.

One day he heard Magda (who was still "mademoiselle" to him) say in her rapid way to one of their friends, "Go quickly and get your sweater. You need it." That command made Trocmé think, Here is a person who cares for others on their own terms, not in order to parade her own virtues, but in order to keep them well. In the many years to follow, he would see how poignantly Magda felt the cold in the bodies of others, and how she would spend much of their lives covering or uncovering children. At the moment, all he could see was that she quite simply cared for others, cared both emotionally and in action.

If there is one image that best symbolizes the relationship of Magda Grilli Trocmé to other people, it is this concern of hers for the physical comfort of people. When you cover somebody with a blanket or a sweater, you are not seeking that person's spiritual salvation; you are concerned only for his or her bodily welfare. And when you cover people, you are allowing their own heat to warm their bodies under that blanket or sweater; you are not intruding on those bodies. You are only permitting them to keep well by their own body heat. In this image of covering lies the essence of Magda's way of caring for others.

They took to walking alone, without their friends from International House, and Magda told him of her early Catholic upbringing in Florence, of her need for liberation from the convent in which she had been placed, and of her not being a Protestant or a Catholic because she believed membership in these or any other sects distracted one from the essentials of the religious life, which had to do with loving one's fellow human beings. And she told him that she was studying to be a social worker so that she could stay close to those essentials.

What might well have angered Trocmé if it had come from the lips of a theologian of the Social Gospel at Union Theological Seminary made his heart leap when it came from Magda's lips. But there were problems. First of all, he had heard her criticize an ascetic friend of theirs for his commitment to poverty, and Trocmé himself was committed to a life working in unions of poor people. She wanted a normal, reasonable married life, and he was not at all sure that their life together could be such a life. Besides, she was on the edge of faith and was critical of all churches. Would she become the wife of a pastor, and if she would, could their life together be happy?

On the morning of April 18, 1926, after standing silently for a while with her near the 125th Street ferry, he asked her to marry him, saying, "I shall be a Protestant pastor, and I want to live a life of poverty. I am a conscientious objector, and that could mean prison as well as all sorts of difficulties." She accepted him, and they sealed their promise to marry with a prayer he uttered aloud as they walked on the rocks of the Palisades.

A few months later, despite the offer of a scholarship from Union Theological Seminary and an invitation from the Rockefellers to spend a year with them in the United States, the couple were standing on the deck of a boat, on their way to France and their wedding.

André had not especially liked America with all its secularism, but Magda had enjoyed the spontaneity of the Americans and had especially enjoyed sharing their freedom, a freedom she had been cheated of all her life in Florence. And so these two stood together on the deck with different emotions, but clinging to each other, as New York disappeared. They would have different emotions and would cling to each other until he died.

5.

Trocmé was faithful to the promise he made when he proposed to Magda: "I shall be a Protestant pastor, and I want to live a life of poverty." Their first parish was in Maubeuge, one of the ugliest industrial cities of northern France, and their parishioners were poor industrial laborers.

Their first child, Nelly, was born there in a burst of blood that almost killed Magda; it was André's brother Francis, a physician, who saved her life. Watching her struggling to survive the shock of losing so much blood so swiftly, André noticed for the first time in his life what he called his "psychological egoism." He realized how in the secret, dark places of his mind he was rejoicing in his own health while he was watching Magda sink toward death. He saw that even at such a time the healthy person, no matter how much he loved the ill one, asked himself what he would do when the other was gone, and even found himself making his own plans for a life without her.

When she recovered, stronger than ever after the birth of her healthy daughter, *she* was not shocked to hear about psychological egoism; she had long ago accepted the fact that people are creatures who desire to stay alive at least as much as they desire to help others stay alive. But André was still a child of the Union of Saint-Quentin. He had to learn with a shock that even people who loved each other could be separate from each other.

There was a deep streak of mysticism in him, a feeling that love could produce a perfect union of two beings. And though he learned about psychological egoism, he never entirely lost this mystical feeling. Magda saw that if he gave way to it his life would be one of ecstasy and not of action; instead of helping others, he would embrace them in ineffectual passion. And so throughout their early married life Magda helped him in his struggle to avoid what he came to call the "abyss of mysticism." Her unfailing common sense held him back from the abyss of useless passion.

In their next parish he came as close to mysticism as he ever would. After a year in Maubeuge, in the fall of 1928, the Trocmés moved to Sin-le-Noble near the Belgian border of northern France, and in this city, which was as ugly as Maubeuge, they spent six years before leaving for Le Chambon.

It was mainly in the kitchens of the homes of poor miners that André Trocmé carried on his ministry in Sin-le-Noble. Other ministers there had spent much time entertaining in the presbytery or enjoying the cultural activities of nearby Douai, but not the Trocmés. Like the Communists, with whom they were always competing for the allegiance of the industrial workers, they loved a life *sans fard* (without cosmetics). They found that life sitting and reading and talking in people's homes, and sometimes praying on their knees together there.

One of the small groups that Trocmé worked with was called the Men's Circle. It was made up entirely of poor people, some of whom were struggling with one of the cruelest enemies of their class, alcoholism. One evening, sitting in a worker's kitchen with the Men's Circle, Trocmé was discussing a book that was very influential at the time, a book that tried to prove that Jesus was a myth invented by Saint Paul. He found himself mustering the arguments and facts he had learned at the University of Paris, but while he was doing so, and, in the process, successfully refuting the book, he also found himself asking the question: "If Jesus really walked upon this earth, why do we keep treating him as if he were a disembodied, impossibly idealistic ethical theory? If he was a real man, then the Sermon on the Mount was made for people on this earth; and if he existed, God has shown us in flesh and blood what goodness is for flesh-and-blood people."

All of this he said calmly to the ten men who were present. He had not planned to say these things, nor had he planned to take any particular action after their talk, but suddenly they found themselves on their knees together. Each made a confession to God of his own weaknesses, as the young people in Saint-Quentin had done, and they all stood up. They found themselves looking at each other with new eyes, without defensiveness, shyness, or pride. They all felt the spirit of God in them, and decided to go right home to bring that extraordinary new awareness to their wives and children.

This was the beginning of what came to be called the "awakening at Sin-le-Noble." In its full intensity it lasted for more than three months, and in the course of it, all the divisions and disputes in the parish disappeared. People became as dear to each other as Jesus was to them. For Trocmé it was "a spiritual springtime. All those things that had formerly been vague, colorless, seen from the outside . . . became suddenly for me living, interesting, inspiring. Each man became inestimably precious in my eyes."

But the "awakening" was not only ecstatic; it involved action. It was not unlike the musical or poetic inspiration that makes some people *productive* geniuses. Such inspiration is not like a mystical trance; it raises people above their ordinary levels of energy, so that, celebrating, they rush out to meet and to change the world around them. Such an inspiration motivated the Hussites in fifteenth-century Czechoslovakia, and the Quakers in seventeenth-century Pennsylvania. They had what Trocmé called a *morale de combat* (an ethic of combat), an active way of living in the world.

The word *good* sometimes carries with it connotations of vapidity. Good children, like good examples, fit neatly and quietly and passively into the patterns others have laid down upon them. The narrator in André Gide's modern novel *The Immoralist* hated "honest folk" because he felt that in them there were "untouched treasures somewhere lying covered up, hidden, smothered by culture and decency and morality."[2] But Trocmé found those treasures then and for the rest of his life only in an ethic of combat. For the rest of his life, imitating the love of Jesus for mankind would be for Trocmé *la grande aventure* (the greatest of adventures). For Trocmé, the "untouched treasures" of creative energy were not smothered by morality; they were revealed by it, and by it alone.

There is no more striking example in Trocmé's early life of the power of this aggressive ethic than the story of Célisse. The story starts during the time of the "awakening." One night the roving squad of the Men's Circle found a man dead drunk, lying in a ditch. The man was Flemish (Sin-le-Noble is only a few miles from Belgium), with a big, square head, the neck of a bull, and vast hands. The circle knew him; like most of the alcoholics in that industrial community, he was destroying his mind and body and brutalizing his family. His wife's skin was gray from suffering, fear, and hunger; it was she above all others who felt the full force of his violent temper and of his cruel neglect. Piece by piece, he had sold almost all of their furniture for drink, and though they had a decent little house on the outskirts of Sin-le-Noble, when Trocmé and the squad entered it, they found the usual home of a drunkard: almost completely empty rooms, and children lying in a corner on a pile of rags, with terror in every line of their faces and bodies.

Under Trocmé's influence, Célisse stopped drinking, visited people, prayed with them, and labored to convince his fellow miners to take the Blue Cross oath against drink. He became the single most effective force in the whole circle for saving people from the hell of drunkenness and anger and remorse. All the power in his mighty body and simple mind was turned toward saving people from destruction, and he blossomed during it all, like a great sunflower.

One day a group from the circle was singing hymns and passing out religious tracts to the miners as they left the pits on their way home for the evening. As was the custom, Trocmé started to make a little speech. Suddenly a small man in a cap joined the group (the "pale guy" of Trocmé's youth was always there, it seems, watching and judging him). In a cutting voice he cried out, "Hey! I know you—you scab! The priests pay you to tell lies. God? There's no God. If there were a God, he would strike me dead right now when I yell 'Shit!' to him. Shut your damned mouth."

Slowly Célisse left his group of friends in the circle. He rolled as he walked, like an athlete, top-heavy with all that bone and muscle. As he walked toward him, the little man in the cap started retreating, yelling all the while. Trocmé had stopped talking, and they were all listening to Célisse and his thundering voice: "What are you saying, you idiot? That there's no God? Repeat that one more time, if you dare, just one more time that there is no God. Have you seen Célisse, eh? Who stopped him from drinking? I'll show you if there's a God!" And Célisse dropped his coat on the road, while the cries of the little man were growing fainter and fainter under the onslaught. But though his cries were getting weaker, they still continued, and Trocmé and the others watched in horror as Célisse raised that fist of his to smite the little man. Only Trocmé's sudden leap saved the blasphemer from being struck to the ground.

"Let me do it," yelled Célisse. "I'll show him that there's a God. These bums, that's the only thing that'll convince them." Trocmé forbade him to strike, and Célisse, baffled, obeyed. On the way back from the exit of the mine, Trocmé tried to explain nonviolence to Célisse, but the pastor's explanation served only to confuse him. He could not understand how nonviolence could be effective. "With such bums there's no other way," he kept muttering all the way home.

At the end of six years in Sin-le-Noble, the Trocmé family left. Jean-Pierre, Jacques, and Daniel had been born there, and they had often been seriously ill in that damp and dirty air. Except for one Belgian family, every one of Trocmé's parishioners had been tubercular at one time or another, and the children were especially vulnerable.

The day in 1934 when the family of six left, Célisse, his wife, and his children appeared all washed and combed in the presbytery. That good-bye, so full of love and sadness, was one of the most memorable and heartbreaking events in Trocmé's life. And one of the reasons it was so painful to Trocmé was that shortly after the Trocmés left, Célisse started to drink again and committed suicide.

For Trocmé, that suicide, horrible as it was, and against his principle of nonviolence, was a statement of Célisse's integrity. With Trocmé gone, he could not restrain his desire to drink. God required him to stop drinking, and so did his friend Trocmé; but he could not stop without Trocmé near. The pastor had not been able to teach him to be nonviolent, but he had inspired him to obey God's commands uncompromisingly. To do this now, he had to leave this life, just as Trocmé had to leave Sin-le-Noble and go to the tiny village of Le Chambon to obey his principles uncompromisingly.

6.

When that refugee called Trocmé the "soul of Le Chambon," she meant that his was the driving force that made Le Chambon the safest place in Europe for refugees. Without knowing it, she was praising him for living up to his *morale de combat,* his aggressive celebration of life. She was saying that he had brought this celebration to Le Chambon, just as he had brought it to Célisse.

But he did not give it to Célisse and he would not give it to Le Chambon in the way that one gives money to the poor or gifts to friends. Trocmé gave his aggressive ethic to them by giving them himself. Aside from the distinction between good and evil, between helping and hurting, the fundamental distinction of that ethic is between giving things and giving oneself. When you give somebody a thing without giving yourself, you degrade both parties by making the receiver utterly passive and by making yourself a benefactor standing there to receive thanks—and even sometimes obedience—as repayment. But when you give yourself, nobody is degraded—in fact, both parties are elevated by a shared joy. When you give yourself, the things you are giving become, to use Trocmé's word, *féconde* (fertile, fruitful). What you give creates new, vigorous life, instead of arrogance on the one hand and passivity on the other.

The giving of oneself is an utterly personal action because each self is a unique person. For Trocmé, giving himself meant "That smile . . . that smile!" as one refugee put it, and it meant a wide embrace because he was such a passionately loving man. One refugee compared his presence to a moving performance of Beethoven's *Eroica* that made you rise to your own highest levels of generosity and joy. This is why Célisse became his best self when he received that gift, and Le Chambon became its best self with his presence.

But there was to be a profound difference between Célisse and Le Chambon: Célisse, with all his physical strength and all his loyalty, was not strong enough to receive Trocmé's vigorous ethic into his own life. He could not live his life without the benediction of Trocmé's physical presence. His suicide showed that he was utterly dependent upon Trocmé. But Magda Trocmé, Édouard

Theis, and the other people of Le Chambon were strong enough to receive his gift of self and go on in their own ways. They could bear to live, on their own, the strenuous ethic he brought when he brought himself to them. They had a power far greater than the power of poor Célisse; they could live a life of moral high adventure on their own, as the last year of the Occupation showed, when Trocmé was fleeing from the Gestapo far away from Le Chambon. It takes more for a Trocmé to succeed than just a Trocmé. It takes also a Le Chambon.

PART TWO

Beginnings—1934–1942

3

The Presbytery and the Flag

1.

A violent mountain storm soaked and shook Le Chambon on the September day in 1934 when the Trocmé family first came to the village. The granite in the buildings, the rock underfoot in the uneven streets, and even the granitic Plateau du Velay upon which Le Chambon stood seemed to be trembling under the thunder, confessing the influence of great power upon them. The rain was beginning to form churning streams in the steeply inclined concave gutters along the sides of the Rue de la Grande Fontaine when the Trocmé family had its first look at the presbytery of Le Chambon.

The fifteenth-century house was gray granite, with a gray stone roof that seemed to weigh it down like a heavy tombstone, so that a whole floor of the house was pressed beneath the surface of the

Rue de la Grande Fontaine. Its closed shutters were gray—not a gray with blue or purple or any other color in it, but a gray from which all life had been washed out, if it had ever been there.

They did not move into the presbytery immediately—workers were inside installing central heating. Instead, they lived at a boardinghouse, which was charging them exorbitant rates. Le Chambon, they noticed, was far from being as hospitable as Protestant parishes in the north of France. One of the reasons for the high rates and the absence of invitations to live for a little while in parishioners' houses was that Le Chambon survived on the tourist industry. It received outsiders strictly on a business basis, charging them the highest rates possible so that the Chambonnais could survive the nine-month-long winter, when villagers kept to their houses because the weather was bitter cold, the streets often too icy to be walked upon on the high, windswept Plateau du Velay, and because there was nothing much to do in the village anyway after the tourists had left. In the winter the village went into hibernation.

For the first few months of his ministry there, Trocmé summarized the spirit of Le Chambon in his autobiographical notes as a village that seemed to be moving toward "death, death, death, and the pastor was entrusted with helping the village die." The people of the village walked around, repeating the wisest maxim they knew: *"Neuf mois d'hiver, trois mois de misères"* ("Nine months of winter, three months of trouble"). The three months of trouble were the three months of summer when the Chambonnais threw themselves into making money from the tourist trade. In the winter they burned for precious heat the *genêt (broom),* the long-fingered bush that had a bitter smell when it was burned indoors, but that brightened the land when it burst into golden blossoms in June. It was the winter with its cold and passivity that dominated their wisdom. The typical reaction of a Chambonnais to some new misery was a shrugged *"Que voulez-vous?"* ("Well, what do you expect? What can you do about it? Nothing, that's what").

The presbytery itself was on the south side of the Rue de la

Grande Fontaine, and its only entrance then was a gate Magda Trocmé calls the "poetic gate." Standing in the street before the gate, you saw on your left a gray bas-relief square picturing a stag standing upon large, stylized palm branches and with an elaborately carved crown the size of the stag hanging above the animal. Some believed that this was the coat of arms of the count of Fay, who used the thick-walled building as a summer home in the fifteenth century, before Protestantism came to France. To the right of the bas-relief was the "poetic gate," which opened into a porch, a heavy roof supported on three sides. The door from the street into the porch was low, but if you stepped back into the street, you saw that it had been cut out of a massive door through which, perhaps, the count's horses and carriages moved in the fifteenth century. Now people used only the smaller door, which opened onto a porch filled with the dusty detritus not of past years—which usually leave a patina that makes dust unimportant —but of recent neglect.

When you stepped through the porch, straight ahead of you was a courtyard overlooking the Lignon River, which flows into the great Loire River. To your right was the granite presbytery, which looked as if it were saying *"Que voulez-vous?"* with an invisible, inward shrug. The entrance to the house itself was shallow, too shallow to protect a stranger knocking at the door from the winds off the Lignon, and it had a wooden door with a large brass knob. Entering, you found yourself in a little, square hallway, facing another door, which opened into the kitchen on your right and the vast dining room on your left.

The kitchen was large and had a big black iron stove with four round holes in its cooking surface. The stove was against the north wall of the house, toward the Rue de la Grande Fontaine, and if you looked along that wall, you could see the rough granite that supported the Rue de la Grande Fontaine. The granite of that wall was usually damp.

The dining room was on the south side of the house. It was very large, and had three windows, two of which faced south onto the

Lignon and the great dead volcano of Le Mézenc far beyond. The windows, having been cut slantedly into the thick stone wall of the house, did not admit much light.

The room was typical of the region: sycamore paneling and ceiling, with a coarse pine floor that made it painful for children to run across in their bare feet because of the splinters. All of the wood was uncarved and unpainted.

In the far west wall of the dining room, next to a fireplace that did not work, there was a little wooden door. It opened into a dark little passage that led to still another door. Standing between the two doors, you were standing in another of the thick stone walls of the old house. Beyond the second door was a dark room, and as you walked into this room, you heard rats pushing things around between the stone wall and the wooden paneling. This cavernous place was to be André Trocmé's office, the spiritual center of Le Chambon during the Nazi occupation of France.

This, with the five bedrooms upstairs and the large beamed attic on the third floor, was the presbytery the Trocmés found in 1934. And it epitomized the spirit of the village then.

But the restless Trocmés did not leave things so. In a little while there was a vast dining table in the middle of the dining room, and the table was covered with a colorful cloth of Basque design whose reds and yellows and blacks enlivened the air around it. There were big geraniums in the three windows, and the windows were decorated with transparent white curtains. In the third window there was a bird cage with a small parakeet in it. From time to time the eldest child, Nelly, who had a talent for the piano and a desire to display it, went to the piano that stood in the corner of the east and south walls of the big room, and played with bold accuracy Mozart's "Turkish March," her favorite. In a rare moment of silence, the big grandfather clock next to the entrance to Trocmé's office ticked loudly, as if to say, "Well, all of this is your frantic way of passing time. This is my steady way."

Inside Trocmé's office there was an Algerian rug of orange,

brown, black, and white; the walls were covered with books and pictures, and he sat in the cold, dark room with huge rabbit-lined slippers on his feet to protect himself from the cold. There was a hole in the wall not far from his desk, and there were rats in the hole who nibbled for long periods of time at a nail there, making small, high-pitched sounds with their teeth on the metal. Sometimes Trocmé stopped his conversation or reading and listened to the rhythms of their sounds. He called them his "musical rats."

But sometimes he drove his body through those two doors and into the dining room "in a holy rage" (as Nelly puts it). Then, in his penetrating baritone voice, he told his children to quiet down —he was thinking, or reading, or conversing. And at other times he left his office and looked quietly at the spectacle of his life in the presbytery. Then he felt gusts of joy rising in his breast, and he would ask himself, "What have I done to be so happy?"

In the course of their stay in the presbytery the Trocmé family made out of a house epitomizing death a place where life could thrive. For four years of their stay, the first four years of the 1940s, this place gave vitality not only to the village around it but to the people who in desperation and need entered that village.

2.

But the transformation of the presbytery and the village took time. Magda transformed the house, but from the beginning it was Trocmé's task—at least he saw it as his task—to transform the village. But how? How could he save this village that was moving toward "death, death, death"?

The inventive Trocmé thought of many enterprises, including a toy factory, but only one seemed able to enliven this village of three thousand inhabitants, two-thirds of whom were peasants living on scattered farms around the center of the village: a school! A secondary school with no ties to the public school system of France, since there was already a public school in Le

Chambon (Darcissac's), and since there was a need for a school leading to the universities; a school that would draw its students and faculty from around the world by virtue of its excellence, by virtue of the fine air of Le Chambon, and by virtue of its freedom from bureaucratic red tape; a school that had a spirit, the spirit of nonviolence, a spirit of internationalism and peace, even though the main function of the curriculum was to be that of preparing teenagers for their *baccalauréats.*

Such a school would keep the tradesmen and artisans of the village occupied during the winter months if it grew to any appreciable size, since the village was small and isolated, and students, faculty, and visitors would have their needs. It would keep the village alive until the tourists came back. But of equal importance, it would help the world outside Le Chambon by sending into it graduates with a deep understanding of the possibility and meaning of nonviolence in a violent world.

By 1938, André Trocmé had found the person he needed to help him start the school: his silent, elephantine friend from University of Paris days, Édouard Theis. Theis had experience as a teacher—he had taught in America, Madagascar, and the Camerоons. And Theis was a conscientious objector. Part of his livelihood would come from the parish, since he would be half-time pastor at the temple, and another part would come from his activities as first director of the school and teacher of French, Latin, and Greek.

In the fall of 1938 the school was tiny, comprising four teachers and eighteen students. There were other teachers, including Mildred Theis and Magda Trocmé, but they were not paid. There was patchwork to be done; they had no teacher of the sciences, and had to send their students to Roger Darcissac's public Boys' School in order to fill this gap. But this was easy to manage, since Darcissac was secretary of the parish and soon a devoted friend of Trocmé. He arranged the whole matter without consulting his superiors in the public school system of France.

Behind the temple there was an annex, divided into sections by very thin walls; here the first language lessons were given, and since one could hear what was happening in the next room about as well as one could hear what was happening in one's own, the place sounded like the Tower of Babel. The temple and its annex were on one side of a busy road, and Darcissac's Boys' School was on the other side of it, facing the temple. The students had to walk around the temple and cross the road in order to study the sciences.

The first few years of the school's history were very difficult for reasons other than convenience, however. From 1938 to 1940, Hitler's Germany was growing in military power and bellicosity, and fear and hatred of the Germans were rising in France. Conscientious objection to war was being attacked more and more passionately as unpatriotic, dangerous sentimentality. And so when they most needed support, at the founding of the school, Trocmé and Theis found themselves deserted by many conscientious objectors, whose patriotism and distrust of Germany were stronger than their love for peace. Many of the pacifists had based their beliefs upon their confidence in the promises of Hitler, but as more and more of those promises were being violated by the Germans, more and more conscientious objectors gave up their nonviolence.

But Trocmé's and Theis's nonviolence had a more stable foundation than that of confidence in the Germans. It was based upon the belief that in the Bible God told us not to kill our fellowman and gave that commandment utter clarity in the life and death of Jesus Christ. And Trocmé had a practical belief, a belief that held out for him a promise of success: he believed that love—that feeling, thinking, and acting as if life is precious beyond all price—would manage to find a way to restrain what his notes call "diabolical forces like Nazism."

What saved the school was not only Trocmé's clear statement of this position to the parish but also the political structure of the Protestant churches of France. Local independence is crucial in that structure. Parishioners elect a presbyterial council, and this council has very great power. It chooses, oversees, and if need be dismisses its pastors with no substantial consultation with regional or national synods. The presbyterial council of Le Chambon endorsed Trocmé's nonviolence and vowed to support him if war came and he was legally designated as a conscientious objector. Since such a designation would make him a violator of the law, such a vow on their part expressed great loyalty to their pastor. Just as Trocmé would not separate himself from Jesus by hating and killing his fellowman, so the Chambonnais would not separate themselves from their energetic minister.

As Hitler grew in power by actions that dropped nations into his hands like ripe plums, the school grew in size. In a few years when somebody asked the stationmaster of the tiny railroad station where the Cévenol School was, he would sweep his arms around and say, "The school is everywhere." There were students, faculty, and classrooms throughout the village. For their classwork students usually stayed in a given room, and their teachers came and went, usually on the run, "like poisoned rats," as the teachers put it. Once Magda Trocmé taught Italian in a large bathroom in the boardinghouse of the Marions, and the annex behind the temple became the center of a complicated network of usually makeshift classrooms.

One of the reasons for the rapid growth of the school was the coming of refugees from Central and Eastern Europe. The love of tyranny and the hatred of the Jew in those places had become clear to them, as clear as the growing military power of Hitler, and so they ran. They ran to many secluded villages in southern France, but in Le Chambon they found not only open doors to homes, but places as teachers and students in the Cévenol School as well. The number of refugees was not great at first (there are no statistics), for the school itself had only forty students, only a few of whom were refugees. The adventure of Le Chambon was barely beginning.

The box-shaped, granite temple of Le Chambon has against its west wall a high wooden pulpit from which the pastor always speaks, after having climbed a staircase to reach it. From that pulpit Theis and Trocmé preached resistance against the hatred, betrayal, and naked destruction that Nazi Germany stood for. They insisted, in those times when some nations were trying to appease Hitler, that a nation, like an individual, must do all it can to resist *le mal* (evil, harmdoing). They attacked the spirit behind the "Keep America out of war" statements that were coming from across the Atlantic; they felt that while evil was being loosed upon the world, neutrality was complicity in that evil.

But their sermons had another aspect: in attacking evil, we must cherish the preciousness of all human life. Our obligation to diminish the evil in the world must begin at home; we must not do evil, must not ourselves do harm. To be against evil is to be against the destruction of human life and against the passions that motivate that destruction.

But the sermons did not propose a neat blueprint for fighting hatred with love. They were not attempts to tell the world or Le Chambon exactly how to overcome Hitler's evil with love. In those last years of the 1930s, the sermons said: Work and look hard for ways, for opportunities to make little moves against destructiveness. The sermons did not tell what those moves should be; they said only that an imitator of Christ must somehow make such moves when the occasion arises. They were preaching an attitude of resistance and of canny, unsentimental watching for opportunities to do something in the spirit of that resistance. Those opportunities soon came.

3.

In June 1940, France fell into the hands of Germany. The armistice of June 22, 1940, divided France into two unequal parts: one usually called the Occupied Zone, and the other usually called, at least by some Frenchmen, the Free Zone. The Occupied Zone, which included Paris and the whole Atlantic coast of

France, was by far the larger and richer part of the new structure. It consisted of northern and western France. The Unoccupied, or Free, Zone comprised all of France south of the Loire River and east of the coastal region occupied by the Germans. Le Chambon was in the Unoccupied Zone. This location in the administrative structure of conquered France was to be of immense importance in the story of Le Chambon.

At the armistice, when Hitler did his little dance of triumph after having taken vengeance against the country that had despoiled and humiliated his own at the end of World War I, France was still united—but only by despair. She was without help under the German boot. Her nearest possible ally was England, but England seemed far away, and tiny, and *next.* As for the United States, it was an unpredictable giant on another planet. Russia was an ally of Germany. On France's southwest border, Spain was another of Hitler's allies. On the southeast border, there was Hitler's friend and partner, Benito Mussolini. To the east was triumphant Germany. The rest was saltwater.

Because of this hopeless situation, resistance to the Nazis or to the Vichy government, which the Germans had permitted to be established in the Unoccupied Zone, was not a task for the self-serving opportunist. Not only was it dangerous to resist, but there seemed to be no future in it. The Thousand-Year Reich that Hitler had promised the Germans seemed to have begun.

One of the students of the 1940s in France, Alexander Werth, summarized the situation of the resistance in this way: "Resistance in the early days of Vichy and of the Occupation was totally disinterested. They resisted because it was, to them, a matter of ordinary self-respect to do so."[3] Resistance, then, was not a "bandwagon" phenomenon. Later, when the Germans were suffering defeats, it would be self-serving, practical to resist them. But in these early days only fools and people of principle resisted. It remains to be seen if the Chambonnais were fools.

The division of France into two zones made an important difference in the types of resistance that would occur in France. In

Paris and in much of the north, German soldiers walked the streets and German administrators were both civil and military leaders. Saying no to these triumphant foreigners was dangerous. They had you by the throat. To resist was to ask for swift punishment—immediate death, internment in disease-ridden French concentration camps, or deportation to forced labor in Germany and almost certain disappearance.

Still, the fear of punishment at the hands of *those* Germans held a certain clarity. There was no doubt about who the enemy was: he was there before you with his German uniform and his German way of acting and of speaking French. Doubt, uncertainty, is a terrible enslaver, like darkness. There was no such enslavement in the Occupied Zone. This is what Sartre meant when he wrote: "We lost all our rights and first of all that of speaking: we were insulted face to face each day, and it was necessary to be still; we were deported en masse, as workers, as Jews, as political prisoners . . . because of all that we were free."[4]

But the "Free" Zone had Marshal Pétain and his National Revolution to darken the mental processes of Frenchmen. By leading the government of Vichy, the eighty-four-year-old hero of Verdun offered the French a French government. In his broadcast announcing the armistice of June 1940, he said, "The government remains free; and France will be administered by Frenchmen only." He was the most conspicuous patriot in France, and certainly her greatest living soldier. He was keeping France from being "Polonized," from being dissected and massacred as Poland had been after her defeat by the Nazis one year before. Why, the good marshal was saving France from humiliation; he was giving her back her economic and political orderliness, and he was restoring her ancient pride. He was the father of a new France that embodied some of the glory of the old monarchical France that had amazed the world from the Middle Ages into the eighteenth century until the disastrous French Revolution. Dared you say no to a man who had restored the essential France in the very teeth of military defeat?

Besides, the French police—mind you, the *French* police—were usually friendly, especially in the early years of the Occupation. They would look away from infractions of the law, and they would even warn their fellow citizens of a coming raid or investigation. How *dare* you turn your hand against these basically kind fellow Frenchmen?

Such questions as these paralyzed much of the power of resistance in the Unoccupied Zone, and it was such a paralysis that Le Chambon had to fight—this and the despair of having been swiftly conquered by the armies of the Thousand-Year Reich. Their success in fighting off that paralysis, in dissipating the fog of doubt, was their first major accomplishment during the Occupation, and this success opened up the way for everything else they did during the four years of the Occupation. Their next accomplishment came shortly after the first.

4.

When the students and faculty of the Cévenol School came back from the summer vacation of 1940, Pétain was firmly in power. France was no longer a republic. Instead of a president, it had a *chef de l'État Français* (chief of the French State). It was forbidden to use the word *République* on stamps and coins. In a move paralleling Hitler's move in Germany at the end of January 1933, Pétain took into his hands the power to summon and dismiss the representatives of the people, the deputies and senators of France. By means of a few "constitutional acts," he brought to France the substance of Hitler's "leadership principle," which vested a plenitude of power and responsibility in the unpredictable will of one man, the leader. The bust or picture of the leader took the place everywhere of the bust of Marianne, the symbol of the French Revolution and of the Rights of Man.

And Pétain used his power in a way similar to the way Hitler was using his power in Germany and elsewhere: to generate solidarity through hatred. Within two weeks of the armistice, the radio, the newspapers, the posters, and the films of France were throbbing with hatred of the Jews (as well as of the Communists, the Freemasons, and even the English). The Jews in particular were soon placed outside the protection of the law in a manner similar to the way the Germans had deprived them of citizenship and had made them "subjects" under the Nuremberg Laws of 1935.

This was the France the students and faculty of the Cévenol School found themselves in during the golden and green fall of 1940. And as if to make the situation clear immediately, the morning before classes began, an order came down from on high: every morning before classes began—and this meant classes in both public and "free" schools, like Cévenol—the flag of France would be raised upon a great pole, and students and faculty would form a circle surrounding it and stand at attention facing the flag; then they would all salute the flag with the stiff-armed, palm-down salute of the Fascists as it waved on high.

Some teachers in the Cévenol School liked the idea. Patriotism and discipline like this would save France from the anarchy that had caused her defeat, they thought. Such a ceremony would help give back to the French what Pétain called the "French soul," unity and purpose.

There were few on the faculty who believed this, however. Some went along with the morning salute with reluctance, waiting for better days. Above all, they wanted to avoid personal discomfort, friction, and conspicuousness. It was such people that all the members of the internment camp at Saint-Paul d'Eyjeaux despised. They were the *attèntistes,* and perhaps in the majority in France in those days. Dante has Virgil describe such people in the Third Canto of the *Inferno:*

> . . . "The dismal company
>
> . . .
>
> Whose lives knew neither praise nor infamy;

...
Who against God rebelled not, nor to Him
Were faithful, but to self alone were true; . . ."

For Trocmé, they were the *"Qu'en dira-t-on?"* ("What will they say?") people, who perhaps thought of themselves as meek and humble Christians, but who were actually cowards who failed to dare, as Christ had dared, to live according to their own consciences.

Trocmé saw the command to salute the flag as an opportunity to begin his resistance to Vichy and Germany. To him, the Fascist salute meant giving up one's own conscience, one's own power to judge human actions as good or evil. He had traveled in Hitler's Germany, and he knew what that salute meant. It meant making some vague and fantastical "national soul" more precious than particular human lives, be they the lives of Aryan true believers or the lives of Jewish victims. No, this was the time to move.

Theis later described the consultation between himself and Trocmé as one in which Trocmé was the mover and shaker. The two discussed the matter briefly and agreed to refuse to give the morning salute in their school. But how should they do it? How should they make that first step?

Roger Darcissac gave them the answer. About fifty feet in front of the temple runs the public road separating the temple from Darcissac's public school; in front of Darcissac's school a low stone wall runs close to and parallel with the road. In his notes Trocmé recalls Darcissac's proposal:

> I shall raise the flag-pole for my school in my own courtyard, but close to the low wall that separates that courtyard from the street. My students will salute the flag by forming a semi-circle inside my courtyard. The students of your school will form, standing in the street, the other half of the circle. They will be looking across the stone wall at the flag. In that way we shall have only one flag ceremony for the two schools.

The Cévenol School would not have to commit itself by raising a flagpole of its own, and could be left free to make its next moves of resistance. Besides, vehicles would be passing on this rather busy road where the students and staff of the school would be standing. The intimate, unglamorous "kitchen struggle" of Le Chambon would thus begin.

Following the plan, Theis, as director of the school, said to the students and faculty, "Those who wish to salute the flag are permitted to stand in the street and do so." For a few weeks, some students formed around one professor who saw the salvation of France in obedience to Pétain. Gradually the number of students in his semicircle grew fewer and fewer. In less than a month, the salute became a weekly ceremony. A few weeks after that it disappeared completely, on both sides of the street.

Apparently, the government was unaware that a few people had refused—at least symbolically—to give up their consciences to the chief of state. But the three thousand people of Le Chambon were fully aware of what had happened. Moreover, they saw that they could resist, that there was a possibility of resisting. For them it was like having a new, beautifully colored bird pointed out, a bird they had not seen before. But now that they saw it, they knew from their own experience that there was such a bird, and it could come back. The bird was the possibility of a person's saying no to indignity and tyranny, and in their recusant Protestant hearts they cherished that bird.

To give the credit for this first move to one person alone—say, to Trocmé—would be to make a serious mistake. He had performed an act of will, a choice; he had started something, as a battery may start a motor, but he had not known when he started it what would be the result of his impulse to resist evil, nor had he known exactly how to bring about that result. It was the excitable Darcissac who had to show him the way for it to happen. Trocmé, as he would do throughout the years of the Occupation, made a choice between two alternatives, and he made it with exuberant confidence that somehow other people would generate the all-important plans that would make that choice effective in the world.

As his sermons showed, he believed that if you choose to resist evil, and you choose this firmly, then ways of carrying out that resistance will open up around you. His kind of originality *generated originality* in others. It did not stifle that originality, the way a dictator using fear and hypnotic charisma stifles the originality of his followers. He generated impulses in them, impulses in the direction of imitating Jesus's love and rebellion, and they went on to make those impulses effective. One will never understand what happened in Le Chambon in those years if one thinks of his leadership as comparable to the august solitude of a glamorous national leader. Trocmé's was an intimate leadership. His life was lived deeply with others—just as his life was lived for others.

Not that his speedy choices were unimportant. They were crucial, and especially important was their timing. Long before he came to Le Chambon, he had learned that a choice had to be made *in time*—not "in due time," not languorously, but in time, *now,* when the hot chain of events had not yet hardened.

In 1921 he had been in the French Army on a mission in Morocco. His job was to help draw up a map of a region near Rabat. To his surprise, one day he was issued a gun and cartridges. Committed as he was to nonviolence, he left the weapon and the bullets in an army depository and joined his lieutenant and fellow soldiers on this, his first mapmaking trip. Once in the desert, the lieutenant noticed that the lanky Trocmé carried no weapon or ammunition, and he asked him why. Trocmé answered that he was a Christian and could not kill; therefore it made no sense for him to carry arms.

The lieutenant took Trocmé into his tent, offered him a cigarette slowly and thoughtfully, and spoke to him about timing. He told the gangling twenty-year-old that he had volunteered for a certain task—mapping a dangerous region of Morocco. And he told him that now they were twenty-five isolated men in an area where robbers and dissidents might attack them at any moment. By giving up his weapon and ammunition he had put the whole group in danger. If the other soldiers had done this, they might all be massacred, but even with one unarmed soldier, their power to protect themselves was considerably weakened.

The lieutenant told Trocmé that his refusal to bear arms had come too late. He had already embarked on a military campaign; he was already committed. He should have refused at the very beginning, when he could have avoided making the march into the desert. With his whole mind and his body he should have made his choice *sooner,* in time.

This conversation in the desert was a turning point in Trocmé's thinking. It taught him that the ethical commandment against killing had to be obeyed as early as possible if it was to be obeyed effectively. It taught him that nonviolence could, in fact, increase violence if it was not chosen in the right way at the right time. It was this lesson that he was using at the very beginning of the Resistance in Le Chambon, and he would use it at every subsequent move in the four years that followed.

He used this lesson a few months after the beginning of the academic year of 1940–1941 when he, Theis, and the rest of the staff of the Cévenol School refused to sign an oath of unconditional loyalty to the venerable chief of the French State. This time he performed his act of refusal with more foresight than he had shown on the occasion of his first act of resistance. His confidence was growing, his timing was getting smoother, and the same was happening to the Chambonnais around him.

These refusals of blind obedience worked. Vichy did not strike, and the people of Le Chambon found themselves discovering not only that the government of France was trying to steal their consciences under the mask of loyalty, but also that they themselves could prevent the theft without being smashed in reprisal. At the end of the first few months of the first academic year of the Occupation, the Chambonnais found themselves, to their mild surprise, living peacefully—and stubbornly—in obedience to the command that is written above the main door to the temple: "Love one another."

4

The Bell and the Empty Buses

1.

On July 31, 1941, less than a year after the salute to the flag had been ordered by Vichy, Pastor Trocmé received a written order from the mayor of Le Chambon instructing him to ring at full tilt the bell of the temple for a quarter of an hour beginning at high noon, August 1. On that day France would be celebrating the anniversary of Pétain's founding of the French Legion of Veterans for Bringing About the National Revolution. He was told that by ringing the bell he would be showing his gratitude to the marshal for having brought discipline and pride into the "soul" of France by establishing that legion.

All members of the legion had taken an oath of allegiance to the marshal himself, and not to any laws coming from the people or approved by them. They were a militant force, the cutting edge of the National Revolution, whose task was to set an example of patriotic enmity against Jews, Communists, and Freemasons. In 1943 they would become the Milice, the dreaded French version of the German SS.

As soon as he received the order to ring the bell, Trocmé went to the custodian of the temple and asked her not to do so. The next day there was silence at noon around the temple. (The little Catholic church of Le Chambon rang its bells, expressing its ancient solidarity with the national government of France.) Trocmé might never have heard what happened at the temple if he had not met the custodian by accident the day after the anniversary in the big, rambling square of Le Chambon.

The custodian was named Amélie. Amélie was a Darbyste, which meant that she was one of the most radical Protestants in France. The Darbystes were followers of a nineteenth-century English preacher named John Darby, who translated the Bible into his own kind of French, and whose preaching drew around him followers who called each other "brother" and "sister" and who rejected even the meager and permissive authority of the Reformed Church of France. After the "desert" of the eighteenth century in France, the French Revolution brought the right to public worship to the country, and under this tolerance many new Protestant sects flourished, the way blood rushes to a lacerated spot when a whipping has stopped. Though the Darbystes were not part of Trocmé's parish, Amélie was a close friend of the Trocmé family, visited the presbytery from time to time, and earned a little money by helping take care of the temple between services. Because they thought that they alone would enter Heaven, the Darbystes pitied others and followed the Ten Commandments all the more zealously for that pity. They would become an important part of the rescue efforts of the people of Le Chambon, constituting as they did almost one-third of the Protestant population of the region. They were staunch protesters against a monolithic state.

When on August 2, 1941, Trocmé stood in the square like a

tree over tiny, sturdy Amélie, two allies were facing each other. Amélie did not begin the conversation; she and her fellow Darbystes were not given to chatter.

"Well, Amélie," the pastor asked, "everything went off well yesterday? No incidents?"

"Everything went off well, Monsieur Tro-ke-mé. There was nothing wrong." Darbyste silence.

But Trocmé insisted, "Come now, Amélie. You were not visited by anyone?"

"Oh, but yes," replied the piping voice with an accent that ended every syllable with a vowel. "There were two ladies from up there"—and a tiny, thick arm swept toward the northern hills of Le Chambon. "You know, those painted ladies who speak proper French." Silence.

Amélie was referring to the people who had come to Le Chambon for the summer from the great cities, and who lived upon the hills in handsome villas. Their cosmetics and their classic French were all that interested Amélie about them.

"Well, then?" asked Trocmé, adopting with reluctance the slow pace of Amélie's speech.

"Well, they came to look for me. And they said, 'You are not ringing the bell, Amélie? It is a national holiday today!' 'The *passe-teur* gave no orders,' I told them. 'Oh, well,' they told me, 'we would really be surprised if he had allowed it to be rung, your pastor! Come, Amélie, hurry! It is noon! And it is an order from the marshal!' "

Part of Amélie's meager repertoire of expressions was a little, crooked smile on one side of her mouth. This smile appeared suddenly as she looked up at the pastor. It was the almost mischievous, youthful smile of protesters who are in complete command of their situation.

"And what did you say?"

"I told them that the bell does not belong to the marshal, but to God. It is rung for God—otherwise it is not rung. Otherwise —no!"

"Bravo! And then what happened?"

"Oh—well, they ordered me to open the big front door, and they told me that they would ring the bell themselves. But I did not want to do that. Then I defended my temple!"

There in the wide square of Le Chambon, Amélie placed herself before Trocmé as she had placed herself before the painted ladies, firmly planted on her feet, her thick arms outspread like the arms of a cross, shivering slightly with enthusiasm and with remembered power, defending "her temple," which was, of course, not *her* temple at all, since the Darbystes worshiped in total independence from the Huguenots.

"And how did it all end, Amélie?"

Again that smile, but now her eyes were round. "Oh, you remember, Monsieur Tro-ke-mé, yesterday at noon it was raining spears, hard. I was under the lintel of the big door. The painted ladies were out in the courtyard. Soon they were dripping wet, and they left."

To Amélie, the incident was not worth bringing to the attention of the pastor. It was only one small part of a long life of protest against compromise with the powerful ones of the earth. But to Trocmé the story was worthy of the Huguenots of old France, who would not renounce their consciences under any threats.

The key word in his interpretation of Amélie's actions is *abjurer.* And this word carries for the Huguenots more than its modern meaning of renouncing one's own commitments. When in 1685 the Edict of Nantes was revoked and the practice of Protestantism became a crime in France once more, the kings of France turned their great power toward making the troublesome "reformers" abjure their Protestantism. They destroyed temples, laid down crushing fines, took children from their parents, and imprisoned, worked to death, or killed adults outright, all for the purpose of making the Protestants abjure their faith and go back to the religion of the majority and of the government. And many of the Protestants who did not join the great waves of refugees fleeing France abjured—out of fear, usually, though sometimes they only appeared to deny their faith in order to survive. To abjure was not necessarily a sign of moral weakness in such terrible times, but it always represented the breaking of an individual's oath under the pressures of the brute force wielded by those in power; and such people as Trocmé and Amélie would never abjure. They insisted on keeping the faith and on obeying their own consciences.

Again, as in the incident of the salute to the flag, it is misleading to think about Trocmé as an isolated leader; he made a decision not to have the bell rung, but without Amélie's particular and Darbyste strength, the "painted ladies" might have rung it. If Amélie had been a compromiser and interpreted Trocmé's order not to ring the bell as having nothing to do with letting the painted ladies ring it, the bell would have been rung. Amélie and the pastor shared a spirit of resistance that was more important than the wording of commands and larger than one man's conscience.

2.

Still, that spirit of resistance was unknown to the authorities outside of the village. In this early period of the Occupation, there was growing in the minds of Trocmé and other leaders of the village a feeling of impatience for an open conflict with the kind of government then ruling France. For people like the ministers, Darcissac, and others in Le Chambon, protest or resistance was not merely a matter of thinking certain private thoughts and performing certain quiet little deeds of a symbolic nature. A full protest involved for them the whole reality of a human being, and part of that reality is public, plainly visible action. Vichy must feel and see their resistance.

On the other hand, a collision could be disastrous for Le Chambon. Elsewhere in France, town leaders were being arrested, and some were being deported to Central Europe, as well as confined in French internment camps. A very strong, provocative move could destroy the leadership of the village, and this would be disastrous not only because it would hinder or even destroy the resistance of Le Chambon but also because it would cripple the expanding rescue efforts of the village. The stream of refugees coming in on the one o'clock afternoon train was growing thicker and thicker, and many of them went straight to the presbytery for necessaries and advice. What would happen to them if the presbytery were emptied and if Darcissac and Theis were arrested? Such thoughts gave the leaders pause.

But events helped solve the problem. In the summer of 1942, a few months before Pétain permitted German troops to occupy the Southern Zone, Vichy decided to send the secretary-general in charge of youth affairs on a formal visit to Le Chambon. The youth of France were very important to Vichy. In July 1940, Pétain had established youth camps in France, and in January 1941, he decreed that all young Frenchmen twenty years of age were to spend eight months in the camps, whose organization was modeled directly on that of the Hitler Youth, down to the Fascist salutes and the indoctrination in zealous patriotism.

And so when in the summer of 1942 Pétain signified that he was sending Georges Lamirand, the minister for youth of Vichy France, to the little village, a very important event indeed appeared on the horizon of that peaceful place: the man in charge of teaching the youth of France to give up their consciences was coming to Le Chambon. As Trocmé puts it in his notes: "For two years we had tried to escape from the hold of the state on our youth. Vichy had first tried to regroup all the young people under the blue shirt of the 'Companions of France': Fascist salute to the flag, bugles, file-pasts, social activities, work camps, worship of the native land and of the marshal. . . ." Now there had to be a confrontation—or a compromise. And one alternative was as dangerous to Le Chambon as the other, though in different ways.

The plan was that there would be a banquet at the YMCA camp

(Camp Joubert) for Lamirand, and then there would be an official march to the sports arena of Le Chambon, where all the representatives of the youth of the region would assemble. After this, there would be a reception in the temple for all the leaders concerned, and finally there would be religious services in the temple. One of the authorities, the head scout of France, assured Trocmé that "It will be magnificent, truly; Lamirand is *très chic* [really splendid]; you'll see!"

At first, things happened that were half-comical, at least in retrospect. Event after event just missed creating open conflict between the people of Le Chambon and Vichy's minister. At the "banquet," Trocmé sat next to Lamirand in the refectory of the YMCA camp. The food was very simple and sparse, and Lamirand, quick and pleasant in his splendid marine-blue uniform (vaguely copied from German uniforms), grinned and said, "It is better this way." He stopped grinning when Nelly Trocmé spilled some soup down the back of that beautiful uniform while trying to serve him. But though he was accustomed to resplendent feasts and flawless, adulatory service, he kept his poise, making the best of a sad situation.

The march through the village on the way to the sports arena had the same tone of a somewhat pathetic lack of energy. The dull gray of the houses was unadorned with flags or bunting, and no one was on the sidewalks. The pastors had expressed their views to the people. Lamirand looked mildly surprised at all the silence and emptiness.

But again, no confrontation; as far as Lamirand and Vichy were concerned, the relationship between Le Chambon and the National Revolution was in as gray an area of human feeling as the gray of those granite buildings. Full awareness of protest meant black and white, not ambiguous gray.

They arrived at the sports field: Lamirand, the prefect and subprefect of the department of Haute-Loire, and the leaders of Le Chambon. Instead of an official reception, Lamirand found himself being pushed around by hundreds of curious children trying to shake the hand of this dignified official with the splendid blue uniform and the soup stain on his back. Again, or still, his quick thinking and poise (he was indeed *très chic*) maintained the fog of good manners. Amidst the handshakes and greetings of the children, he was charmed by so much "spontaneity."

It was a rather cool day up there on the plateau, even though it was mid-August. A Chambonnais gave a short talk on the thirteenth chapter of Saint Paul's Epistle to the Romans. It had to do with the respect due to authorities. There is no copy of the speech, but the thirteenth chapter opens: "Let every person be subject to the governing authorities. . . . Therefore he who resists the authorities resists what God has appointed." And it urges Christians to do their civic duties, like paying taxes and honoring those to whom official honor is due. But it goes on to say: "Owe no one anything, except to love one another; for he who loves his neighbor has fulfilled the law. The commandments, 'You shall not kill, You shall not steal, You shall not covet,' and any other commandment, are summed up in this sentence, 'You shall love your neighbor as yourself.' Love does not do wrong to a neighbor." To anyone who knew the chapter—and the people of Le Chambon knew it well—the ethic of neighborly love demanded not a bitter confrontation with the government but a perfunctory, minimal respect for the "governing authorities," with a firm but quiet hint that there are limits to that respect, limits set by the commandment not to do wrong to a neighbor.

Lamirand was puzzled. He had prepared a long speech, but he replied with only a few words. Still, grayness.

In the temple a visiting Swiss pastor gave a sermon on the theme of obedience to the state when the state does not try to force the people to violate the laws of God and does not try to violate that one law that sums them all up: "You shall love your neighbor as yourself." In his notes, Trocmé says that the visiting pastor got out of it very well. Trocmé passed Lamirand a hymnal as they sat in the big gray hall of the temple, and, uncomfortably, the official tried to sing. Afterward they rose and walked out into the front court that faced the road beyond, where once the salute to the flag had been grayly protested.

And then it happened. A dozen of the older students of the Cévenol School, some of them future theologians, walked up to the handsome Lamirand, handed him a written document, and begged for an acknowledgment of it on the spot. In his notes, Trocmé quotes the message:

> Mr. Minister:
>
> We have learned of the frightening scenes which took place three weeks ago in Paris, where the French police, on orders of the occupying power, arrested in their homes all the Jewish families in Paris to hold them in the Vel d'Hiv. The fathers were torn from their families and sent to Germany. The children torn from their mothers, who underwent the same fate as their husbands. Knowing by experience that the decrees of the occupying power are, with brief delay, imposed on Unoccupied France, where they are presented as spontaneous decisions of the head of the French government, we are afraid that the measures of deportation of the Jews will soon be applied in the Southern Zone.
>
> We feel obliged to tell you that there are among us a certain number of Jews. But, we make no distinction between Jews and non-Jews. It is contrary to the Gospel teaching.
>
> If our comrades, whose only fault is to be born in another religion, received the order to let themselves be deported, or even examined, they would disobey the orders received, and we would try to hide them as best we could.

Black and white. The maneuvering between the two obligations to be "subject to the governing authorities" and to "love your neighbor as yourself" was past. The moment had come for the people of Le Chambon to pass their ethical judgment publicly, unequivocally, but without hatred or violence.

At last, the colorful Lamirand wilted and turned pale, *chic* as he had been, and said, "These questions are not my affair. Speak to the prefect of your department." And he hurried into his auto, away from these smoldering-eyed Protestants.

Prefect Bach was angry, and he knew exactly where to turn to express his anger. Instead of addressing the young men, he said, "Pastor Trocmé, this day should be a day of national harmony. You sow division."

Trocmé did not shirk the responsibility of having planned the confrontation. He said, "It cannot be a question of national harmony when our brothers are threatened with deportation."

Prefect Bach replied, "It is true that I have already received orders and that I shall put them into effect. Foreign Jews who live in the Haute-Loire are not your brothers. They do not belong to your church, nor to your country! Besides, it is not a question of deportation."

Trocmé asked, "What, then, is it a question of?"

"My information comes from the marshal himself. And the marshal does not lie! The Führer is an intelligent man. Just as the English have created a Zionist center in Palestine, the Führer has ordered the regrouping of all European Jews in Poland. There they will have land and houses. They will lead a life that is suitable for them, and they will cease to corrupt the West. In a few days my people will come to examine the Jews living in Le Chambon."

Trocmé replied, "We do not know what a Jew is. We know only men."

Then, at this moment of full awareness on the part of both sides, the prefect tried to impale Trocmé and Le Chambon on the other horn of the dilemma that was draining the power of the French to tell right from wrong: now that the mental fog of doubt was gone, he threatened Trocmé, even as those lucid ones in the Occupied Zone were being threatened by sheer force. "Monsieur Trocmé," he said deliberately, "you would do well to take care. If you are not prudent, it is you whom I shall be obliged to have deported. To the good listener, warning." And the prefect left.

At this time, the existence of extermination camps such as Auschwitz and Maidanek, where millions of Jews and others were being humiliated, tortured, and killed, was unknown to the people of Le Chambon, including Trocmé. As a matter of fact, the extermination of the Jews (as well as of the Gypsies) was going on, but all Trocmé and the people of Le Chambon knew was that "it is evil to deliver a brother who has entrusted himself to us. That we would not consent to."

Trocmé did not know much beyond this, but he *realized* what was at stake. The *Nacht und Nebel* (Night and Fog) policy of the Germans regarding the death camps was successful. And part of the fog the Nazis had created around those camps were stories like that of a Polish Zionist state for Jews. As Trocmé accurately put it in his notes, "Many French let themselves be deceived in 1942."

This is a psychologically penetrating statement. The Chambonnais under Trocmé, on the other hand, would not let themselves be deceived. Trocmé knew enough about Nazism and cared enough about its victims to realize that what the Germans were doing—whatever it was—was not for the good of the Jews. Perhaps he did not *know* more about Nazism than many other Frenchmen—Hitler's anti-Semitism was no secret in Europe—but he *cared* enough about its victims to realize what giving the Jews to the Germans meant for the Jews. That caring had to do in part with Saint John's commandment to love one another, but it also had to do with stubbornness, if you will, fortitude, a refusal to abjure—this crucial word again—a commitment. The Chambonnais were committed to sheltering the Jews. They would abide by that commitment despite all threats from the governing authorities.

There was no fog for them because they cared enough to see and to act and to be firm. The *attentistes,* the waiting ones who played bridge, perhaps, with friends, the collaborators with Vichy who expedited the deportation of foreign Jews from France, and the Nazis did not care for the victims of Nazism. That mixture of lucid knowledge, awareness of the pain of others, and stubborn decision dissipated for the Chambonnais the Night and Fog that inhabited the minds of so many people in Europe, and the world at large, in 1942.

The students in their letter to the minister (in all likelihood Trocmé wrote it, or at least had a very heavy hand in its writing, but again, his leadership was so intimate that this is hard to clear up) mentioned the incident at the Vélodrome d'Hiver in Paris. They opened their remarks with a description of the *rafle* (roundup) of Parisian Jews that had taken place there three weeks before Lamirand's visit to Le Chambon. On July 16–17, 1942, the foreign and stateless Jews of Paris were arrested by the French police under orders from the Germans.

Approximately 28,000 Jews were taken. Single persons without children were sent to the concentration camp at Drancy, about three miles northeast of Paris, where living conditions were brutalizing. Families with children were sent on the Paris city buses to the Vélodrome d'Hiver, the sports arena south of the Eiffel Tower, where they were held for eight days. Some Jews realized that they were going to their doom, and committed suicide before the police arrived. One doctor killed himself and his family with injections of strychnine.

The Vel d'Hiv, as it is usually called, is a large arena covered by a glass roof painted blue. At that time it was lit by naked bulbs. After a short while, the lavatories stopped working on those hot July days, and the stench in the closed arena became insupportable. At first, people performed their private functions sitting on the filth; but after a while this became impossible, and men, women, and children had to perform those functions in public. There was no fresh air, and the dust, with the immense stench, made the air difficult to breathe. There was no water, either for drinking or washing. Some women ran screaming through the arena begging for death, and there were ten successful suicides.

Through it all, the French police expressed almost total indifference, even to the 4,051 children. The crying of unfed, unwashed children and the screams of their despairing mothers filled the stinking air of the arena. In the fullest published account of those terrible eight days, *Betrayal at the Vel d'Hiv,* by Claude Lévy and Paul Tillard, an eyewitness tells one of his memories:

"I shall never forget one little girl. She was sick. With her eyes glued to my face, she was begging me to ask the soldiers to let her go. She had been a good girl all year; surely she didn't deserve to be put in prison."[5] Not a single child survived the roundup and consequent deportation to Auschwitz. They became part of the more than one million Jewish children murdered by the Nazis.

It is impossible to be sure of exactly how much of all this Trocmé and the Chambonnais knew a month after the event. But it is plain that the man to whom they presented their letter, Georges Lamirand, knew a very great deal about it. He was head of the General Secretariat for Youth and had sent some uniformed young people into the Vel d'Hiv to do some cleaning and help carry stretchers bearing some of those who were unable to walk. One newspaper, *Au Pilori,* had called this assistance "shameful" and "hateful servitude." And it is possible that this story, published on July 23, weeks before the Lamirand visit to Le Chambon, was known to the Chambonnais.

But one thing is certain: unlike the 9,000 French police, and others, the Chambonnais cared about what had happened, and cared enough not only to feel pain at the suffering of the victims but also enough to make inferences about future events. They cared enough to think and plan: "Knowing by experience that the decrees of the occupying power are, with brief delay, imposed on Unoccupied France, where they are presented as spontaneous decisions of the head of the French government, we are afraid that the . . . deportation of the Jews will soon be applied in the Southern Zone." Further, they announced that they would not be deceived by the "Free Zone" label applied to southern France. They cared enough to watch events closely and to see that there was a pattern: first the German masters up north would take oppressive measures, often by using French police as a screen, and then, as if on their own initiative, the "French" government in the south would take similar measures.

I have mentioned that it is impossible to draw a hard, clean line between what Trocmé felt and thought and did and what the rest of the Chambonnais felt and thought and did. He was immersed in Le Chambon. But *his* caring was important to theirs, and his caring came from a lifetime of concern. From the loss of his mother on, death had been a part of his life, and he had hated it, without hating the killer. As a young man in Belgium, he himself had been a refugee. These and other experiences had prepared him to care; they had sensitized and informed him, so that his caring was fed by knowledge, and his knowledge was spurred by caring.

3.

Two weeks after Lamirand's visit, Vichy struck. It was a Saturday night late in the summer of 1942 when at first a few automobiles and later some khaki-colored buses surrounded by police motorcyclists entered the marketplace of the village. As soon as they arrived, they summoned Pastor André Trocmé to the town hall, which is close by the square.

Trocmé found himself facing the chief of police of the department of Haute-Loire, a very important official of the Vichy government. (The man was not Silvani, who was later to arrest Trocmé and the other leaders of the Resistance in Le Chambon.) The time for official politeness was past, and the chief of police was direct. According to Trocmé's notes, he said:

> "Pastor, we know in detail the suspect activities to which you are devoted. You are hiding in this commune a certain number of Jews, whose names I know. I have an order to lead these people to the prefecture for a control. [Trocmé injects here: "He was lying; it was a question of deportation."] This must be done with order. You are therefore going to give me the list of these persons and of their addresses, and you will advise them to be on their good behavior, so that they should not try to flee."

Nobody was *chic* now. Trocmé replied that he did not know the names of these people. He was telling the truth. They had false identity cards, and Trocmé did not want to know their real names.

Trocmé went on, "But even if I had such a list, I would not pass it on to you. These people have come here seeking aid and protection from the Protestants of this region. I am their pastor, their shepherd. It is not the role of a shepherd to betray the sheep confided to his keeping."

The chief of police was becoming angry. He replied, "What I said to you is not advice, but an order. If you oppose authority, it is you who will be arrested and deported. I hold you responsible for this unacceptable resistance to the laws of your country. Besides" —and here Trocmé records the fact that he had a villainous laugh— "your resistance is useless. You do not know the means which modern police employ: motorcycles, automobiles, radios—and we know where your protégés are hiding!"

The conversation was ended. Trocmé walked away and turned into the little street on which stands the presbytery. First he called the Boy Scouts of Le Chambon into his cavernous office, and then he sent each of them to certain outlying farms to warn the Jews to flee into the woods during the night. The whole operation (which Trocmé called the "disappearance of the Jews") had been carefully worked out immediately after Lamirand's visit, and it scattered Jews not only into the thick woods around Le Chambon but into the department of Ardèche, which lies to the east of the Plateau du Velay upon which Le Chambon stands.

Saturday night was a busy night in the little village. Something went wrong with the lighting system, and so in the darkness one could see various figures, usually Boy Scouts and Bible class leaders (the backbone of Le Chambon's Resistance, the young *responsables*), crossing the town to get to and from the farms and to and from the various houses in the village itself that sheltered the refugees. Under a starlit night, it was as if ghosts were purposefully making their respective ways through the square and the streets while the police waited for their ultimatum to expire, sleeping upon straw.

When Sunday morning came, the two pastors expected to see the police waiting in the square to arrest them. But they were not there, and during the service in the temple no police were visible. According to one account, Trocmé's own, in his notes, the police came later and waited near their buses in the square on alert, late into Sunday afternoon. According to another account, which has been confirmed by Pastor Theis and by Trocmé's daughter, Nelly, the police were far from idle during the temple service; they conducted a careful search of the houses of Le Chambon and of the outlying farms. One thing is sure: the temple was full that Sunday morning, and tense. There were even people standing in the front hallway, just inside the big front door that faces Darcissac's Boys' School.

In the large, rectangular chamber of the temple, the two pastors spoke from the high pulpit against the west wall, and what they presented to the people was a declaration that urged them to obey God rather than man when there is a conflict between the commandments of the government and the commandments of the Bible.

They may have invoked the chapters in the Old Testament concerned with safeguarding the persecuted in "cities of refuge" (Numbers 35:9–31; Joshua 20:1–9; Deuteronomy 19:1–13). In Deuteronomy 19:7–10 it is written: "Therefore I command you, you shall set apart three cities . . . then you shall add three other cities to these three, lest innocent blood be shed in your land which the Lord your God gives you as an inheritance, and the guilt of any bloodshed be upon you." In the Bible it is people guilty of what we now call involuntary manslaughter who are protected in the cities of refuge until they can be brought to fair trial. But the Jews were being persecuted not because of any crime, voluntary or involuntary, but only because they were Jews. Trocmé and Theis must have felt that these modern Jews were all the more deserving of refuge because of their utter innocence of any crime or even any charge. In any case, the ministers were concerned only with safeguarding the weak from the unjust hatred of their pursuers, and they were willing to ignore other aspects of the "city of refuge" passages in the Bible.

But one part of those passages they were not willing to ignore was the statement of the Lord that if the innocent are slain in a city of refuge, the guilt of that bloodshed will be upon those who committed themselves to sheltering them in their city. The ministers—and the people they led—believed that they had a duty to protect the refugees, a duty that, if violated, would bring condemnation upon them as betrayers of a trust and of a commitment.

Many of the obligations laid down in the Bible involve avoiding doing harm. The Ten Commandments, for example, lay down such negative obligations: you shall not kill; you shall not commit adultery; you shall not steal; you shall not bear false witness against your neighbor; and so on. Ordinarily, people have a strong obligation only to avoid doing harm themselves; they are not usually obliged to go out of their way to do anything that will *prevent others* from hating, hurting, or deceiving. It is usually enough if they simply sit quietly within the limits laid down by the "you shall not's" and do nothing to violate those limits. People are not often *required* to help. We are not often obliged to obey the Ten Commandments and do *more.*

But the spirit of the passages about the cities of refuge makes the prevention of harmdoing, the prevention of injustice, a requirement, a heavy obligation upon those who live as regular inhabitants of those cities. They must both refuse to do harm themselves and act to prevent others from doing harm, as if they were all being commanded to be the Good Samaritan of Luke 10:30–37, and as if *this* was what Jesus meant when he said, immediately before the Good Samaritan passage, "You shall love the Lord your God with all your heart, and with all your strength, and with all your mind; and your neighbor as yourself."

It was this strenuous, this extraordinary obligation that Theis and Trocmé expressed to the people in the big gray church. The love they preached was not simply adoration; nor was it simply a love of moral purity, of keeping one's own hands clean of evil. It was not a love of private ecstasy or a private retreat from evil. It was an active, dangerous love that brought help to those who needed it most.

When the sermon was over, there was great emotion in the church. In pain and fear, the parishioners said *adieu* to their pastors as they all made their way to the street that divides the grounds of the temple from the grounds of the Boys' School. But when they looked around, they saw no police in the street waiting to arrest the leaders.

After midday, the reason for this became clear. During the service, the village council, seated between machine-gun-armed police, had received from the police chief of the Haute-Loire department instructions to sign an "Appeal to the Jewish Refugees," which asked Jews to present themselves at the town hall for a *recensement* (census). When he saw this "appeal," Trocmé thought, A fine census this is, with two or three buses standing in the market square, ready to take them away for deportation! It later became clear that a summer resident of Le Chambon had consented to be the author of this treacherous appeal, which began, "Measures indisputably understandable. . . ."

But no Jews presented themselves at the town hall. They had been warned. And by the end of the afternoon the police had gone to work. This is how Trocmé describes their actions in his notes:

> They searched first the houses in the village and of the closely surrounding country, calling for identity papers from everyone, opening cupboards, going down into the cellars, climbing to the attics, knocking on walls to see whether they were hollow. They showed themselves polite, sometimes rough—but they found no one.

On Monday they fanned out to the farms in the surrounding countryside. Actually, such a two-part search was foolish, since the refugees in the surrounding farms had a whole night to escape.

One Austrian Jew named Stekler was arrested. He sat in one of the buses, surrounded by several policemen, and the villagers smiled at him as they passed through the square and stared at the empty buses—several policemen with one lone prisoner to be guarded! Jean-Pierre, the oldest son of the Trocmés, in whose life compassion was one of the most powerful forces, gave Stekler his last piece of rationed imitation chocolate. Others brought more presents, and soon the quiet little man had a pile of gifts beside him almost as big as himself. He was later released, having only two grandparents who practiced the Jewish religion—at this time, being "half-Jewish" was legally the same as being classified *non-Juif* (non-Jewish).

One other person was arrested, and there are conflicting reports about whether she was deported to the death camps or whether she was saved before her train entered Germany. Édouard Theis believes that she was not saved.

The police remained in Le Chambon for three weeks, firing up their motorcycles early in the morning, trying to surprise any Jews who might have tired of life in hiding and gone back to the homes that had been sheltering them. They found no more Jews. The dogs on the farms of Le Chambon warn their masters—and the whole countryside—of new arrivals, barking at even the most distant sounds. As Trocmé put it in his notes, "For centuries, in these wooded countrysides where silence dominates, people know all that happens."

4.

There are two stories that are now legends in Le Chambon—though Trocmé assures us in his notes that they are true—regarding that strange *rafle* of the summer of 1942. They remind one of the half-comical beginnings of the Resistance in Le Chambon, but they have a meaning of some importance to understanding this later stage in the Resistance.

A police lieutenant, handsome in his new Vichy uniform, just happened to be allowed to walk on some rotten planks over a cesspool on one of the farms being searched by the police. Apparently, no one thought of warning him. There was an ill-smelling rescue, and the wilted lieutenant was comforted with coffee.

After the first day of the *rafle,* the people of Le Chambon lost some of their nervousness and regained their sense of humor. When asked about Jews, they stared back in astonishment and said, "What would Jews be doing here? Have you seen any Jews yet? They say they have crooked noses!"

The Chambonnais were becoming seasoned citizens of a city of refuge.

The second story has a different meaning. Étienne Grand, the son of the mayor of Le Chambon, was reading with his back against a tree when a gendarme stopped at some distance from him. Trocmé tells the story in his notes:

> "Psst, psst," said the policeman.
>
> Étienne continued to read.
>
> "Hep, over there," called the gendarme more loudly. ["Hep" was a derogatory word addressed to Jews.]
>
> Étienne raised his eyes. The gendarme made a strange motion: with a gesture he indicated to Étienne that he ought to "get the hell out of there." Étienne did not understand. The embarrassed policeman approached: "Go away," he said to Étienne Grand. "I haven't seen you!"
>
> Étienne said, "You haven't seen me? But what does all this mean? What are you trying to say?"
>
> "I'm looking for Jews. If you don't leave, I shall be forced to arrest you."
>
> "But I'm not Jewish," exclaimed Étienne, in whom the light was dawning. "I am the mayor's son!"
>
> "Ah, you aren't one of them?" cried the relieved Pandora. "So much the better—I do not like the task they make us do."

As the Resistance in Le Chambon developed, a curious phenomenon was taking place there: many of the Vichy police were being "converted" (as Trocmé puts it in his notes) to helping the Chambonnais and their Jews. Even as the official policy of Vichy toward Le Chambon and the Jews was hardening, *individuals* among the police and the bureaucrats of Vichy were more and more frequently resisting their orders to catch or hurt people who had done no visible harm to anyone. They found themselves helping those who were trying to save these innocent, driven creatures. Caring was infectious.

When, halfway through the Occupation, Nazis came down to occupy southern France, and when the Gestapo was putting iron in the souls of Vichy policemen, there would often be a telephone call before a *rafle.* A mysterious voice would warn one of the Trocmés: "Attention! Attention! Tomorrow morning!" And there would be a click as the receiver was swiftly replaced. Magda Trocmé says that the voice probably came from the nearby city of Le Puy, where the main German garrison for the region was stationed. But wherever it came from, it was usually accurate, and it would trigger the "disappearance of the Jews."

Flesh-and-blood contacts with people who would not hate and who would not harm people communicated something of what Theis and Trocmé used to call the "mystery of love" to some rather hard human beings. We shall see how this communication reached into the leadership of the German troops at Le Puy.

5·

With the *rafle* of the summer of 1942, the Resistance of Le Chambon lost much of its symbolism—the saluting of flags, the ringing of bells, the giving of oaths dissolved as important issues for the Chambonnais. And even humor and pathos all but disappeared. What was left was the one activity that made Le Chambon a village of refuge: the saving of innocent lives. The days of symbolic refusal—important as they were—were past. From the summer of 1942 on, the Chambonnais confined themselves to the most dangerous commitment of all: sheltering and saving the bitterest enemies of the Nazis, the Jewish refugees.

Shortly after the *rafle,* in November 1942, the Germans moved down to occupy southern France. From this time on, Vichy police were closely supervised by the Gestapo, the secret police of Hitler, and the danger of giving aid to Jews became far greater than ever. In the face of this, the Chambonnais concentrated on making Le Chambon a city of refuge.

On a high, wide plateau, surrounded by mountains and extinct volcanoes, with the peak of the volcano Le Lizieux to the west, stands the village of Le Chambon-sur-Lignon.

The entire Trocmé family in Le Chambon before 1940. In front of André and Magda stand their children: (from the left) Daniel, Jacques, Jean-Pierre, and Nelly.

For many refugees the "poetic gate" of the presbytery (above) was the real entrance into the village of refuge. To the left of the door is the ancient coat of arms of the village. Once through the poetic gate they came upon the "shallow door" of the presbytery (below).

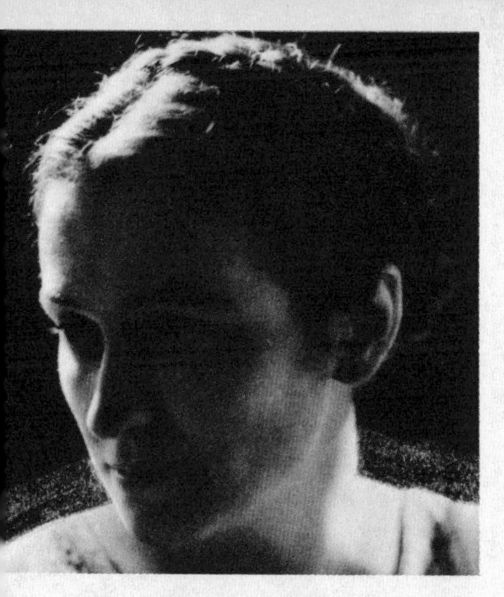

Magda Trocmé at the beginning of the war, and Daniel Trocmé, second cousin of André Trocmé.

Refugees arriving by the one o'clock afternoon train alighted here, at the railroad station, which served both Le Chambon and her sister village Le Mazet.

Two views of the Protestant Temple of Le Chambon: the main entrance (above, left) with the inscribed legend "Love one another," and the bell in the tower which the Darbyste Amélie refused to ring (above, right).

Inside the Temple André Trocmé preaches from the *chaire* at the west wall, early in the Occupation.

A typical farmhouse in the countryside of Le Chambon surrounded by the woods that offered swift escape to refugees during surprise "roundups."

The House of the Rocks, where young male refugees were housed and fed, and where Daniel Trocmé and his wards were arrested in the summer of 1943.

Two photographs from the internment camp at Saint-Paul d'Eyjeaux in 1943: (above) from the left, Édouard Theis, Roger Darcissac and André Trocmé, and (below) Édouard and André conducting services for their fellow inmates.

Édouard Theis in Madame Eyraud's home, 1977.

Roger Darcissac in his study, 1977.

André Trocmé in Geneva in the 1960s.

The tomb of André Trocmé in the cemetery behind the Temple of Le Chambon. His son, Daniel, who died after the war while residing in the United States, is buried under the same headstone. To the left, Madeleine (Manou) Barraud, who was accidentally shot late in the Occupation, is buried. To the right, Jean-Pierre, who was found dead in the presbytery late in the Occupation, is buried. The marker reads: "The peace of God is beyond all understanding."

PART THREE

Help—1940-1944

5

Burned Shoes and the Quakers

1.

The winter following the conquest of France—the winter of 1940–1941—was one of the most ferocious in the history of modern France. In Le Chambon the snow was piled high against the gray walls and buildings, and *la burle* seemed never to stop its whirling. In the granite presbytery Magda Trocmé was putting small pieces of wood and dried genêts into the kitchen stove to keep the heat as high as possible without wasting fuel. The big black stove stood against the wall facing the Rue de la Grande Fontaine, the crooked street along which her husband would later walk on his way to an internment camp.

Concentrating—as she always did—upon the details of what

she was doing, the calculation of just the right number of bits of fuel to put in the stove, she was slightly startled to hear a knock at the outer door of the presbytery. Someone had come through the "poetic gate" that opened onto the Rue de la Grande Fontaine, and was standing in the shallow doorway of the presbytery itself, standing in the wind and snow blowing off the Lignon River. She closed the stove, went through the dark little hallway, and opened the outer door.

There before her, only the front of her body protected from the cold, stood a woman shawled in pure snow. Under her shawl her clothes, though once thick, had been whipped thin by the wind and the snow of that terrible winter. But her face had been whipped even thinner by events; she was visibly frightened, and was half-ready to step back, trembling with fright and cold. The first thing that Magda Trocmé recalls seeing was the hunger in that face and in those dark eyes.

Here was the first refugee from the Nazis to come to the presbytery door. This is the way Magda Trocmé started her description of the incident in a conversation with me thirty-six years later:

"A German woman knocked at my door. It was in the evening, and she said she was a German Jew, coming from northern France, that she was in danger, and that she had heard that in Le Chambon somebody could help her. Could she come into my house? I said, 'Naturally, come in, and come in.' Lots of snow. She had a little pair of shoes, nothing. . . ."

There was a big wooden table in the middle of the kitchen, and, after asking her to take off her shawl and coat, Magda had the woman sit at the side of the table facing the stove, where she gave her something to eat. They started to talk. The pastor's wife spoke German, and she soon learned a story that was to become familiar. The woman had run away from Hitler's Germany because of the increasingly vicious racial laws, and had found herself in the Occupied Zone, where the lives of foreign Jews were in constant danger; and so she had kept running until she crossed the demarcation line and entered the Free Zone, where there were fewer German troops and less danger.

Magda Trocmé is not given to long, ruminative chats. When she talks, she does so swiftly, with a breathless, heavy voice whose pace and tone seem to be saying, "Come, now, let's get this talking over with and *do* what we're supposed to be doing." She has powerful arms that are frequently being used to move things and to push herself away from tables and up from chairs so that she can cope with human physical needs instead of the "nonsense" she often finds in conversations between idle people. Accordingly, her first reaction, after seeing that the woman was fed and comfortable, was to get up and go for help. Idle compassion was as alien to Magda Trocmé as idle talk.

Standing, she calmed the woman with a few words, urged her to warm herself at length near the stove, and to put her soaking-wet shoes, for a little while, in the oven attached to the stove. Then she put on her own shawl, pushed open the kitchen door, went through the little hallway, out into the bitterly cold courtyard, through the "poetic gate," and into the streets of Le Chambon. She walked west toward the square and the town hall.

The snow and wind were incidental to her; mainly she was thinking that the woman had to have *papers,* especially identification papers. There were frequent surprise checks for papers and frequent roundups, even this early in the Occupation. Without papers, the woman was in danger of deportation back to Germany and Hitler.

She walked up the few steps of the village hall and into the mayor's office. She remembers that she was confident that the papers would be forthcoming; after all, there were Frenchmen in this village, not Nazi racists. Almost as a matter of course, she told the mayor about her German refugee, expecting him to sit down with her and help her make plans for the welfare of the refugee.

And so the mayor's response surprised her. "What?" he said. "Do you dare to endanger this whole village for the sake of one foreigner? Will you save one woman and destroy us all? How dare you suggest such a thing to me? I am responsible for the welfare of this French village. Get her out of Le Chambon tomorrow morning, no later."

She did not argue; she simply arose and went back into the snow. As she went down the hill toward the presbytery, she analyzed the situation. She had gone to the authorities and revealed to them the presence of an illegal, unregistered refugee in her home. The fact that the refugee was in the home of the Protestant minister of a Protestant village not only exposed the refugee to action by those authorities; it seemed to invite such action, since the presbytery was the single most conspicuous home in the village, and the duties of the authorities were clear. The mayor, Frenchman though he was, was still under the command, ultimately, of Hitler. The mayor could justify his action not only by pointing to the photograph of the handsome, eighty-four-year-old Marshal Pétain, who had ordered the "surrender on demand" of foreign refugees, but also he could and did justify it by pointing to his own role as leader and protector of the village of Le Chambon-sur-Lignon. For him, moral obligations held only in the realm of "one of us": native Frenchmen; they did not apply to "one of them": foreigners, German Jews. For him, it was "our people, our lives"; it was the trust laid upon us by our leaders that determined what was good and what was evil.

All of this had something to do with the mental fog in southern France, the incapacity to resist a French leader above whom the French saw only vaguely, in a cloud of doubts, the Nazi Führer. But the mayor had emphasized the other source of paralysis in France: fear of reprisals, fear that a whole village might become an object of suspicion, and, because of the Nazi conquerors in great numbers above the demarcation line, even a target for destruction.

But Magda Trocmé did not dwell upon these matters during that walk down from the town hall. Her main thought was that the refugee was now in great danger because she was known to the mayor. For her to stay in Le Chambon against the orders of the official leader of the village, *and with his full knowledge of her presence and location,* was absurd. If Magda allowed her to stay in the presbytery, she would, in effect, be dooming her to capture.

And so Magda Trocmé had to get her out of Le Chambon. Like others in the Unoccupied Zone, she had a "line" to other people who opposed Pétain's measures. Even in those early days of the Occupation, many resisters knew of others who could be trusted in an emergency; and Magda Trocmé knew of a certain Catholic family who were enemies of the National Revolution and who were courageous and compassionate persons. She would send the refugee to them.

Probably because of the separation of Catholics from Protestants in France (I have met some French Catholic families who have never knowingly met a French Protestant), she did not know that there was growing near Lyons, not more than a hundred miles from Le Chambon, a powerful Catholic network whose main task was to save Jewish refugees: the *Témoignage Chrétien* (Christian Witness), the organization of the courageous Father Chaîllet. In any case, at this early stage in the Resistance, people had only the names of scattered individuals on their "lines."

When she came back into the kitchen, she saw the woman still sitting before the stove, shoeless. Swiftly Magda ran to the oven beside the stove and opened the oven door. The shoes were still in the oven, burned black. The two women gasped.

For the refugee to leave by the next morning, she had to have shoes. But Le Chambon was a poor village, especially in the winter months when there were no tourists. And Le Chambon was poorer than ever now, what with this extremely severe weather and the difficulty of finding supplies after the supply lines to the cities had been smashed in the sudden, overwhelming defeat of France. It was especially difficult to find shoes, since the adults in the village could not buy shoes legally and could not afford to buy them on the black market; only children could buy them, one pair a year, on their birthdays and with a coupon. This was one of the reasons why the awkward wooden shoes of old were being used more and more of late; there was wood enough in the forests in and around Le Chambon, and there were still some woodworkers. But shoes for long walks were almost an impossibility to find for adults of modest means. André Trocmé —who had agreed completely with every move she had made on behalf of the refugee—once got a pair of leather shoes for his big feet only because somebody about his size had died in the not-too-distant city of Saint-Étienne, and a friend had arranged for him to receive them.

Magda Trocmé is not one to work hard before a crowd of idle onlookers; she is like *la burle*—busy, but also making everything around her busy as well. That night there was a network of shoe-seekers in the village, and she was pushing through the snow in the middle of the network. Even in that harsh winter she brought back the shoes before the night was far gone.

The next morning, the woman left. Later Magda was to reflect, "She was a Jew. What did she think of a Christian community, walking in the snow without knowing where she was going?"

It is true that Magda Trocmé had sent her to another family, but who was to help the woman get there? And exactly what would she find there? For the rest of the Occupation, Magda Trocmé and all the other people of Le Chambon would know that from the point of view of the refugee, turning somebody away from one's door is not simply a refusal to help; it is an *act of harmdoing.* Whatever one's excuses for not taking a refugee in, from the point of view of that refugee, your closed door is an instrument of harmdoing, and your closing it does harm.

A while later—while her husband was away—another Jewish refugee appeared at the presbytery door. Magda Trocmé had a fresh idea. Vichy is Vichy. Understood. But what about Jews helping their fellow Jews? At this time, the French wife of an influential French rabbi was staying in Le Chambon to escape the rigors of the Occupied Zone where she and her family were permanent residents. Magda went to visit her and asked her to help with this German Jew. The answer she received struck the iron of egotistic human reality even more deeply into her soul than had the answer of the mayor: "A German Jew? But it is because of the foreign Jews that our French Jews are persecuted. They are responsible for our worries and difficulties." Again that hard line between "one of us" and "others"; again the idea that moral obligation has to do only with "one of us," not "one of them."

These two events were important in the history of the presbytery and of the refugees in Le Chambon. They were important not only because these two refugees were the first of hundreds who would come to the presbytery door, and not only because these two events showed how narrow the domain of love was to some people, but also because they helped the Trocmés realize the concrete meaning of the "city of refuge" passage in Deuteronomy 19:10: "lest innocent blood be shed in your land . . . and so the guilt of bloodshed be upon you."

And there was another lesson, a very practical one, that the Trocmés and Le Chambon learned from these events: they must *conceal* from the authorities and from unsympathetic citizens any help they were giving refugees. To reveal that help would be to betray the refugees, to put them in harm's way. Either conceal them or harm them—those were the alternatives.

But in Le Chambon in the beginning of the 1940s, concealment meant lying—lying both by omission and by commission. It meant not conveying to the authorities any of the legally required information about new foreigners in Le Chambon, and it meant making false identity and ration cards for the refugees so that they could survive in Vichy France. It meant, for example, changing the name Kohn to the good old French name Colin so that the refugee could have the life-giving identity and ration cards to protect against roundups, when identity cards were usually checked, and to protect against hunger, since the basic foods were rationed. Such cards made it unnecessary to report a new foreign refugee to the mayor—only a Frenchman with, perhaps, an accent had come to town.

But for Magda and the other Chambonnais, the making of counterfeit cards was not simply a matter of practicality. It raised profound moral problems. To this day, Magda remembers her reaction to hearing about the making of the first counterfeit card. During that first winter of the Occupation, Theis came into the presbytery and said to her, "I have just made a false card for Monsieur Lévy. It is the only way to save his life." She remembers the horror she felt at that moment: duplicity, for any purpose, was simply wrong. She and the other leaders knew that ration cards were as important as identity cards—the Chambonnais were so poor that they could not share their food with refugees and hope to survive themselves. Nonetheless, none of those leaders became reconciled to making counterfeit cards, though they made many of them in the course of the Occupation. Even now, Magda finds her integrity diminished when she thinks of those cards. She is still sad over what she calls "our lost candor."

How, then, could they lie and violate one of the commandments given to Moses? Theis and Trocmé saw that deep in Christianity is the belief that man is never ethically pure—in this world he finds himself sinning no matter what his intentions are. The best he can do is acknowledge and lessen his sins. Such a view is part of Judaism, too. In 1972, Magda Trocmé went to Israel to participate in the ceremony awarding her husband—posthumously—the Medal of Righteousness. Part of the ceremony involved planting a tree in memory of André Trocmé (there is a tree in Israel for every person who has received the Medal of Righteousness). During the ceremony, one of the speakers said, "The righteous are not exempt from evil." Magda remembers the sentence word for word. The righteous must often pay a price for their righteousness: their own ethical purity.

She is aware of these depths, but they do not comfort her. She still feels anguish for the children of Le Chambon who had to unlearn lying after the war, and who could, perhaps, never again be able to understand the importance of simply telling the truth. But usually when she says this, she suddenly straightens up her body, with typical abruptness and vigor, and adds, "Ah! Never mind! Jews were running all over the place after a while, and we had to help them quickly. We had no time to engage in deep debates. We had to help them—or let them die, perhaps—and in order to help them, unfortunately we had to lie."

But her daughter, Nelly, points out that the children, as far as Nelly could see, never had the problem of unlearning lying. She remembers the children, among them herself, seeing the situation with the clear eyes of youth. She remembers their seeing that people were being helped in a desperate situation by these lies. And the children were convinced that what was happening in the homes of Le Chambon was right, simply right.

What the children saw was what the rest of the Chambonnais saw: the *necessity* to help that shivering Jew standing there in your door, and the necessity not to betray him or her to harmdoers. In this way of life the children were raised, and—at least according to Nelly—they did not feel their parents to be guilty of any wrongdoing.

There were many women in Le Chambon whose homes were the scenes of events like those in Magda Trocmé's kitchen. There was, for instance, round-faced, sparkling-eyed Madame Eyraud, whose husband was *très chic* in the violent Maquis, despite her own nonviolence. When I asked her why she found it necessary to let those refugees into her house, dragging after them all those dangers and problems, including the necessity of lying to the authorities, she could never fully understand what I was getting at. Her big, round eyes stopped sparkling in that happy face, and she said, "Look. Look. Who else would have taken care of them if we didn't? They needed our help, and they needed it *then.*" For her, and for me under the joyous spell she casts over anybody she smiles upon, the spade was turned by hitting against a deep rock: there are no deeper issues than the issue of *people needing help then.*

The fact is that the Chambonnais were as candid, as truthful with the authorities as they could have been without betraying the refugees. Trocmé was perfectly willing, as were the other Chambonnais, to tell the authorities that there were refugees in Le Chambon. As a matter of fact, they felt that it was their duty to do so, and the letter to Lamirand says this outright: "We feel obliged to tell you that there are among us a certain number of Jews." The spirit of Le Chambon in those years was a strange combination of candor and concealment, of a yearning for truth and of a commitment to secrecy. They were as open as love permits in a terrible time.

2.

Magda's words to her first refugee, "Naturally, come in, and come in," were part of an ethical action. Ethics, especially the ethics of crisis, of life and death, deals with the lives and deaths of particular human beings. In the context of a life-and-death ethic, refusals like the refusal to ring the bell of the temple are ethical only insofar as Amélie and André Trocmé were ethical, only insofar as particular human beings were involved in their deeds. Such refusals made the difference between life and death for a given individual only indirectly—they helped Le Chambon to resist Vichy and the Nazis, and so helped develop the spirit that would save particular human lives. But it is one thing to resist a government and its National Revolution; it is another to face a shivering, terrified Jew on your doorstep. Life-and-death ethics has to do with hurting and helping individual human beings. It has to do with betraying, torturing, humiliating, killing them, and with helping them.

When that first German Jew appeared in the doorway of the presbytery, ethics became incarnate in Le Chambon for Magda Trocmé. Only then were two individual human beings involved: one in danger, and one being asked to help. This was what Magda Trocmé meant when she said to me, "Helping Jews was more important than resisting Vichy and the Nazis."

But how did Le Chambon become a place where so much help was being given that the police had to come with their automobiles and buses in order to try to stop it? How did Le Chambon

become a village of refuge? This much is certain: in the course of the first two years of the Occupation, Le Chambon became the safest place for Jews in Europe. How did a life-and-death ethic become incarnate across the whole commune of Le Chambon?

3.

Even in normal winters, the mistral rushes down the Rhone River valley from the icy Alps at speeds between thirty and fifty miles an hour, especially from December through March. But the winter of 1940–1941 was an especially bitter one, and the "masterly wind" (the literal meaning of the old Provençal word *maestral*) struck the great port city of Marseilles with damp, cold, unremitting force. There was much rain, and there was even snow in that southern city. The mistral cut through the thin, wet clothing of the many refugees wandering through the streets, and it penetrated the windows and doors of many houses. Influenza was common, not only among the refugees who were going from door to door looking for help in obtaining the life-giving papers that would permit them to leave France; it was common even among those who had lived through many Marseilles mistrals.

Later in the winter of Magda Trocmé's encounter with the refugee, André Trocmé walked out of the Gare Saint-Charles, the main railroad station of Marseilles, stiffly, as was his manner with that painful back. He lumbered down the many steps just outside the main entrance to the station, and stepped onto the Boulevard d'Athènes. Not far away was the beige-and-black four-story building at 29 Boulevard d'Athènes. The two middle stories of this building were the offices of the American Friends Service Committee. Trocmé went in.

He had just obtained permission to make this visit from the presbyterial council of his church. At a recent meeting he had told his parishioners that he felt the need to help the many refugees who were pouring into southern France from Central and Eastern

Europe. Many of them had been put into the terrible internment camps that had sprung up across southern France. The camps were practically destitute of clothing, food, and medical supplies, and the filth was maddening. All his life he had kept close to Jesus Christ by helping those who were suffering. He had urged his parish to let him continue this work in the part of France that needed him most, and they had consented. Moreover, they were ready to give him supplies and money for the refugees.

He had come to the Quakers because he was convinced that he could best show his love for Jesus and for his fellowmen by working with the Friends, who were already bringing desperately needed supplies and consolation to people in internment camps like the ones at Gurs and Argèles. With this in mind, he walked into one of the stark little offices and introduced himself to Burns Chalmers, who was responsible for many of the Quakers' activities on behalf of the inmates of the camps in southern France.

Trocmé knew that the Quakers had long ago grown proficient at carving out for themselves an ethical space within which they could move with comparative freedom. He knew that what they did in that space was turn all their psychological and financial resources toward diminishing suffering and death. I do not know how much André Trocmé knew about the Quakers' activities in the nineteenth century on behalf of the slaves and freedmen of the United States, but he knew a great deal about what they had done in Germany after World War I. He knew that from 1918 on they had fed about a million and a half German children with what the Germans still called *Quaker Speise* (Quaker food). The Germans remembered even as late as 1940 that the Quakers were no nation's enemies and no nation's friends. The Germans—and André Trocmé—knew that their only enemies were suffering and killing, and each human being was their friend.

Trocmé knew also that the Friends had carved out another ethical space in France. They had been trusted enough by the Vichy government to be allowed to care for the many kinds of sufferers in southern France: victims of war, starving children in schools, and prisoners in the camps.

The two had never met, but Chalmers had heard of Trocmé as a nonviolent leader whose prestige and influence in southern France were very great. It is important to see that for Chalmers, as for other Quakers, the power to lead people was of central importance. Though the Quakers did not take sides in any national or political struggles, they had to be sure that the nations in which they worked would not hinder their efforts. Freedom to act within their ethical space was essential to their efficacy; without it, they could not be of service to those in need; without it, they would have in their hands only the worthless coin of simple compassion. And so they had to be protected by powerful leaders native to the country in which they worked. Trocmé, after only a half-dozen years in the south of France, was widely recognized not only in his own department of Haute-Loire, but in other departments, including neighboring Ardèche, as an important religious leader.

Ever since the seventeenth century when the Englishman George Fox founded the Religious Society of Friends, they have had two main functions: witness and action. They have witnessed to—that is, embodied in their lives—the idea that a person should walk cheerfully over the land, seeing God in every man. Their lives have expressed the feeling that every human being has a precious spark of God in him, and that therefore he or she is as valuable as God. And, being so valuable, every human being has a right to be treated truthfully, lovingly, and, if in need, helpfully. That precious spark of God is the "inner light."

The other function of the Society of Friends is simply service, help. From the very beginning, Quakers have been active in personal and social crises. Though they have refused certain kinds of actions, like worshiping in state churches, taking oaths, and bearing arms in wartime, they have insisted on performing other kinds of actions; and the purpose of these actions has always been the alleviation of suffering.

Quakers seem to have a certain oscillatory rhythm in their lives. On the one hand, there is a streak of mysticism in them, a deep involvement with the invisible inner light; but on the other hand, they plunge into horrific situations and live as if action and not feeling were the main reality in the world. In my talks with Chalmers thirty-six years after his meeting with Trocmé, he told me that the Friends "instinctively stress the deed, rather than the thought or faith processes behind the deed." When they see a victim of natural or man-made tortures, the inner light seems to lose its importance to them. When they see suffering, they do not invoke invisible forces; rather they throw themselves into the center of a disaster area and give visible help.

All of this was part of Chalmers's thinking as he met the Huguenot about whom he had heard so much. This Frenchman with three hundred years of French Protestantism behind him and this American with the international Quaker experience behind him had the same ethical commitment: to treat human life as something beyond all price.

Chalmers is a tall, thin man who often questions himself and his motives painfully, with "concern," to use the Quaker term. Trocmé was also tall, though somewhat heavier, with a driving, even aggressive, manner. Chalmers listens and feels, and probes with long, thin fingers into the workings of events; Trocmé listened and felt and probed, but had to do so quickly, almost impatiently, and always *en route* (on the march). Chalmers is inward; Trocmé was militant, with those massive shoulders of his always rising by a sheer effort of will above the pain in his back. But these differences between a painfully inward man and a soldier did not make their first meeting difficult. On the contrary, their differences made the men fascinating to each other. Chalmers was at this time centrally concerned with persecution in Occupied France. According to Chalmers, centuries of persecution had given the Huguenots, and Trocmé in particular, what Chalmers calls "a sturdy quality." Being a minority had helped make them clear-cut in their thinking and firm in their convictions.

Having been tested by adversity, they had kept themselves alive by remaining lucid and unshakable. The Psalmist in Psalm 26 had said to God: "Prove me, O Lord, and try me; test my heart and mind." History had done this for the Huguenots.

In his urgent, rapid voice, Trocmé presented to Chalmers his plan to enter the internment camps with the Quakers. In his first presentation of his plan, he won Chalmers's Quaker-practical heart by avoiding the name of God and by refusing to discuss the ethics of helping the inmates. His words and his intensity appealed immensely to Chalmers, who saw before him a volcanic man whose detailed plans flowed from him with tremendous rapidity and vivid imagery. But behind the volcanic eruptions of this man, at the center of the life of the volcano, was the simple force that Chalmers always looked for in any person asking to work with the Quakers. Thirty-six years after the event, Chalmers remembered that the force behind everything Trocmé said was this: "He cared intensely for persons."

But impressed as he was by the man and by the force that drove the man, Chalmers was not ready to rush into helping Trocmé. Chalmers's office was almost as busy as the Gare Saint-Charles, and almost as full of refugees bearing the overwhelming burden of fear and confusion. Besides, Chalmers was not a man who, like some executives, could engage in two major tasks at once. The power of Trocmé seemed to demand Chalmers's entire concerned attention, but so did the eyes of each refugee. As for the refugees, the head of the Toulouse Quaker group had recently said, "The sum of human misery is enormously increasing on all sides." And Chalmers saw the truth of this not only in the crowded streets of Marseilles and the other cities and ports of southern France but also in the face of each Central or Eastern European who walked into his office with an invisible but nonetheless terribly real sword of Damocles—arrest and deportation—hanging by a single hair over his head.

Engulfed with human suffering and its rightful demands upon him as a Friend, Chalmers told Trocmé that their conversations could not continue in Marseilles. Because of the centrality of its location between Le Chambon and Marseilles, the two decided to meet in Nimes about once a month in order to talk without interruption about how Trocmé could help the refugees.

Near the Maison Carrée in Nimes there was a very simple restaurant. In it, at a side table, they discussed not goals—they were in complete agreement about goals—but *how* Trocmé could help "overcome evil with good," to use the words of Paul's Epistle to the Romans.

Both men were especially concerned with children. They wanted to give the children of the refugees a strong feeling and a solid knowledge that there were human beings *outside their own family* who cared for them. Only by *showing* them that human beings could help strangers could they give those children hope and a basis for living moral lives of their own. The most obvious way of doing this was to alleviate the suffering of those children.

But there were different ways of doing this. How should Trocmé proceed? Chalmers was convinced of the importance of bringing supplies and encouragement to families in the camps. In fact, he had done much to alleviate the suffering of families in the camps by bringing material and philosophical encouragement to them.

However, as the conversations in Nimes progressed, both Trocmé and Chalmers came to believe that such visits to the camps were not the best way for Trocmé himself to overcome evil with good as far as the children were concerned. Trocmé was a man as much *with* others as he was *for* others, and so his decisions were almost always deeply intertwined with the decisions of his allies. The momentous decision to make Le Chambon a refuge for children was in this respect typical of Trocmé. To this day, I do not know who put the idea of sheltering children in Le Chambon into Trocmé's mind, Chalmers or Trocmé himself. The two were in such intimate communication in that Nimes restaurant that it is difficult, perhaps impossible, to tell who influenced whom the most.

According to Trocmé in his autobiographical notes, it was Burns Chalmers who had the most active role in making the final decision: "Burns did not advise me to go live in the camps [there were some French Protestants and Catholics who were then doing so]: 'We already have various organizations for doing this. . . . *But* you tell me you come from a mountain village, where one can still enjoy a certain security.' " Chalmers had seen in Le Chambon the crucial condition for effective action on behalf of the children, the weakest of the weak: security, freedom, within a given ethical space where goodness could overcome evil without hindrance from the outside world. A village surrounded by rugged mountains and on a high plateau that was difficult to reach.

According to Trocmé, Chalmers went on to say that the Quakers were now trying to deliver to inmates in the camps as many medical certificates as possible; these medical certificates declared the person named to be incapable of labor and therefore ineligible for deportation or other kinds of forced labor. At this time, at the very beginning of the Occupation, "deportation" did not mean destruction in the extermination camps; rather, it meant *Zwangsarbeit* (forced labor) in Germany. Chalmers told Trocmé that when they could not save the father of a family by delivering such a certificate of medical exemption, they tried to save the mother. At this point in Trocmé's notes, the words of Chalmers are put in capital letters: "IF THE PARENTS ARE NONETHELESS DEPORTED, WE TAKE CHARGE OF THE CHILDREN, AND WE ARRANGE THAT THEY BE BOARDED OUTSIDE OF THE CAMPS." But Chalmers went on to say (still according to Trocmé) it was very difficult to find French communities that would agree to run the terrible risk of receiving such dangerous guests. Finally, in this account that makes Chalmers the driving force behind the idea of Le Chambon as a refuge, Chalmers posed the key question to the spiritual leader of Le Chambon: "Do you wish to be that community?" And Trocmé's notes go on, casting Trocmé himself in the role of a questioner, even an objector: " 'But these children, it will be necessary to board them, to nourish them and to educate them. Who will take charge of this?' 'Find some houses and some monitors,' said Burns. 'The Quakers and the Fellowship of Reconciliation [an international group created after World War I for spreading the philosophy of nonviolence and helping victims of natural and man-made disasters] will support you financially.' "

This is Trocmé's account of the Nimes conversations, an account written, apparently from rather slight notes, about thirty years after the event. Chalmers disagrees with it. According to him, the idea to shelter the children in Le Chambon emerged in much the same way a "sense of the meeting" emerges in a Quaker meeting, by a common effort. He does not remember ever making such pointed suggestions—out of the blue, as it were—to the dynamic Trocmé. In fact, Chalmers insists that he could not have made some of the remarks Trocmé ascribes to him. For instance, the remark about the Quakers financing, or helping to finance, the operation in Le Chambon—Chalmers was not the top official in Marseilles, and he had no right to commit Quaker funds on his own initiative. He had a position of leadership in the American Friends Service Committee, especially as far as helping refugees was concerned, but a financial commitment was beyond his power to make. In fact, Chalmers was almost certain, even over a gap of thirty-six years, that he did not make a commitment to help the Chambonnais.

Nevertheless, the decision and the plan to make Le Chambon a safe place primarily for children emerged in the course of these conversations in the little restaurant near the Maison Carrée in Nimes. Chalmers's task in the implementation of that plan was to obtain funds for a house, and supplies; Trocmé's was to obtain this house, appoint its monitor, and arrange for the distribution of supplies to the children of parents who had been deported to forced labor in Germany. Both gave their word that they would do all that was possible to accomplish these things throughout the whole period—which then looked endless—of the Nazi occupation of France. Actually, they would both go beyond their commitments and would develop more than one house of refuge in Le Chambon.

4·

This account of the decision to make Le Chambon a place of refuge for children must have one more element in it if it is to help us understand what the Chambonnais did during the four years of the Occupation. And that element is an understanding of why Burns Chalmers did not accept Trocmé's original plan to work with the Quakers in the internment camps. We have noticed that one of the reasons for this was that there were already various people working in the camps, either with the Quakers or independently of them. But this is a slender excuse for turning Trocmé away from the camps; with all the refugees and their many problems, one could always use one more helper, especially someone like André Trocmé.

Chalmers had a better reason for wanting Trocmé to work in his village. He told Trocmé: "It is difficult to find French communities which will agree to run the risk of receiving guests . . . [who] . . . are so dangerous." Not many communities had the leadership or the will to put their lives, and possibly the very existence of the community itself, in danger. Most communities were too concerned with their own self-interest, their own self-preservation, to undertake such a task. Moreover, most French communities were not willing to disturb their already difficult personal lives by accepting foreigners into their very *foyers* (hearths), their kitchens, bedrooms, and dining rooms. But Le Chambon would be willing; Trocmé could vouch for that.

There was yet another reason for Trocmé's going back to Le Chambon instead of working in the camps. This reason was André Trocmé himself, his personality and his commitments. In my conversations with Chalmers, this reason emerges as by far the most important one.

Chalmers saw in Trocmé first of all an effective leader, especially an effective spiritual and ethical leader. But for Chalmers, who has an eye and a heart for the individuality of people like no man I have ever met, it was the humane warmth of Trocmé, his openness to fresh feelings and new ideas, that made him the man to lead a place of refuge whose purpose was not only to save the lives of children but also to set an example of overcoming evil with good. Trocmé was always growing, always becoming stronger while he gave strength to those around him—he grew *with* them. He was the kind of man who could give children hope because he would not seem to be giving them anything except his own powerfully loving self.

Working in the camps involved working under a bureaucracy in strict obedience to narrowly defined commands. Trocmé's originality and his passion required wider and more flexible limits than those set in the camps. Such a man needed spiritual elbow-room in order to move his own way. Chalmers told me that he saw Trocmé as a "major figure" and as the leader of a little world wherein "there was no limit to what might be possible in terms of the reclamation of persons, in terms of giving new hope to the outlook of a person." And he told me that even after more than a third of a century, he still felt grateful for having met such a man.

6

The Spirit of the Presbytery

1.

After an evening of meetings with the *responsables* of his parish, André Trocmé would return to the presbytery. He would walk through the wooden door that opened into the outer porch, stride across the little courtyard to the shallow main entrance to the presbytery itself, push open the door, enter a dark little hallway, go through one last door on his left, and find himself in the dining room with his four turbulent children.

"Papa! Papa!" they would cry, not because he brought them gifts—they were too poor for that—but because he brought them himself. Two of the three boys would seize his legs, one each, and stand on his feet so that he could make an "elephant walk" with them; the third boy, older and bigger than the other two, would wrap his arms around his father's waist and stand with his head under his father's armpit. And his daughter, the eldest child, would throw her arms around his neck and almost strangle him while he tugged and mussed her hair with his free hand.

Amidst the chaos, Magda would come from the kitchen through the little square hallway into the dining room. She would come for no reason except to watch her husband standing there, as Trocmé describes it, "like a tree covered with branches and fruit." But no real tree ever used its branches so; he would squeeze his children in his arms until someone would complain; this they called the "lemon-squeeze" game.

Sometimes, if he was too tired or too full of pain to play such games with them, they would sit around him while he told them a newly minted episode in the unending story of the "Little Beast." The Little Beast had very special powers. She could become as small or as large as she wished; and she could become as light or as heavy as she wished. And she could do more: she could become very heavy *while being* very small, so that her great weight could do mischief in a person's pocket or pocketbook. The Little Beast embodied *le bien* (goodness), and she always took the side of somebody too weak to defend himself. Her purpose was not to hurt or punish the evil ones as much as it was to save the underdogs from harm. Her main task was to affirm life, not to destroy evildoers. And she always won out over *le mal* (evil), always succeeded. The end of each story was a celebration of the joy of living and of being saved from destruction. Even now, Nelly remembers those celebrations of the victory of goodness over evil —how *glad* the family felt, and how much she learned about good and evil from her father's stories of the Little Beast.

In the course of the Occupation, such moments became rarer and rarer. There would be more and more refugees waiting in their dark clothes there in the dining room; Papa would come back later and later, tireder and tireder, and more and more often, all he could do was make his way across the dining room and into his somber office, trying to recover his strength and keep clear his vision of the many activities that were happening in Le Chambon.

Usually there were twelve people at the dining table: the six Trocmés, one of the two permanent refugee boarders (it was usually Monsieur Colin—Madame Berthe Grünhut was in the kitchen cooking), a young woman who helped Magda with the children, and four adolescent boarders, who were students at the Cévenol School. The income the Trocmés received from boarding these four students was, as Nelly now puts it, "an income very much needed for our survival. We were very poor."

The Trocmés were exciting to each other, but with the refugees coming and going in the presbytery with their horrific stories, and the large number of different people at the table, the four years of the Occupation were the most exciting period in the lives of the children. In fact, as Nelly puts it, the children found themselves "spoiled by all this excitement." Her friends envied her this continuing excitation, and after the liberation of France life in Le Chambon seemed boring to her, compared to those years.

But excitement is one thing, and a deeply satisfying family life can be another. There was no family living room in the presbytery, no place where the Trocmés could be together alone and at ease. Family chats were rare, and they took place by invitation: the minister would ask his family to join him in his study, the only luxurious place in the house because it was the only place with a rug. These invitations would come on Mondays, Trocmé's official day off. But more and more Mondays became workdays for him as the Occupation went on and the number of refugees in the commune increased; and the invitations were never frequent enough for the children.

If her friends envied Nelly the excitement at the dining table, she envied them the intimacy of their living rooms. The chairs in the house were high-back, wicker-bottom wooden chairs that did not invite repose. And aside from the two permanent refugee guests and the four students in the house, there were almost always strangers in the dining room, preoccupied with their own terrible problems. Without ever putting their feelings into words, the children felt that their parents were not available to them. They felt that too much of their parents' lives was being given over to this one task, the helping of refugees, a task that was not that big in the lives of the children. Once Nelly said that she would not marry a minister—she did not want to be dedicated to a cause that was larger and more important than her family.

But the children never resented the regular intrusions into their privacy. Young as they were, they knew that there was something that had to be done—people needed their help. They also knew that their family had to be the ones to do it. And they knew that what there was to be done had to be accomplished with all one's heart and mind or it could not be realized at all. The children gladly ran errands for the refugees, and Nelly remembers that for her and her brothers their father was a hero and their mother a heroine.

Nelly was a regular churchgoer and a convinced Protestant. But her younger brother Jacques had a more stormy relationship with the temple and the Protestant religion. In January 1943, a few weeks before his father's arrest with Theis and Darcissac, and a few months after the Germans had moved down to occupy southern France, an event occurred that shook and cracked the foundations of his allegiance to the Protestant church.

One evening there appeared in the dining room one of the top leaders of the Reformed Church of France, the church in which his father was a pastor. The boy was awed by the presence in his house of what seemed to him to be "the equivalent of the pope" (as Jacques Trocmé puts it now). And his awe was intensified to the point of trembling by the fact that the man had a large mustache that swept down on either side of his mouth, and that reminded the boy of the ancient French hero Vercingetorix, who had heroically tried to defend Gaul against Julius Caesar. Here in his dining room were a pope and a Vercingetorix wrapped into one!

The religious leader told the boy's father that he had something of great importance to discuss with him, and the two men went through the two doors into his father's office, closing the inner door behind them. A ten-year-old who could seize the clothes of German soldiers bathing in the Lignon and lead them a merry chase was not to be daunted by fear of a reprimand, especially if he could hear his father talking with one of the great temporal and spiritual leaders of France about something very important. And so "Jacot" sat himself down beside the closed door of his father's office, in the dark little hallway that had been carved in the wall between the dining room and that office. And he listened.

VERCINGETORIX: What I want to say is this: you must stop helping refugees.

TROCMÉ: Do you realize what you are asking? These people, especially the Jews, are in very great danger. If we do not shelter them or take them across the mountains to Switzerland, they may well die.

VERCINGETORIX: What you are doing is endangering the very existence not only of this village but of the Protestant church of France! You must stop helping them.

TROCMÉ: if we stop, many of them will starve to death, or die of exposure, or be deported and killed. We cannot stop.

VERCINGETORIX: You must stop. The marshal will take care of them. He will see to it that they are not hurt.

TROCMÉ: No.

When Jacot rushed through the doorway to the dining room, he heard them still arguing passionately, with longer speeches from Vercingetorix and shorter speeches from his father. More than a third of a century after this event, Jacques Trocmé remembers this conversation with anguish. Listening to the religious leader, he lost his trust in church organizations and in churches themselves. From that time on he did all he could to avoid going to catechism or to the temple. As he puts it now, "I could not then and I still cannot stomach the ideas of the leader and of the church he represented." One Sunday morning in Le Chambon, I asked him to join me in the temple. He would not, and he explained his refusal by telling me this story.

Shortly after overhearing the conversation, he tried to talk to his father about it, but Trocmé shook his head and refused. When Jacot asked his mother about it, all she would say was, "The leader is a good man, Jacot. You do not understand his position." Later the boy learned that the leader became more understanding of Trocmé as the Occupation wore on.

Later, Jacques Trocmé learned about the sequel to the conversation he had overheard. His father decided to submit his resignation to the presbyterial council of Le Chambon. Their reply was unanimous on two counts: a flat refusal to accept that resignation, and strong, clear encouragement that he should go on helping Jews and other refugees—against German orders, against Vichy laws, and against his own church's wishes.

In later years, Jacques Trocmé learned that this whole series of events (the reprimand from the leaders of the church, the refusal on the part of André Trocmé to change his ways, Trocmé's offer of his resignation to the presbyterial council of Le Chambon, and their refusal to accept his resignation) had all happened before. In 1939 the leaders of the Reformed Church told André Trocmé that his policy of conscientious objection to violence was in error doctrinally and could do the church harm, especially in time of war. Four years before the episode of the saving of refugees, Trocmé went to the presbyterial council of Le Chambon and asked them to accept his resignation. Then, too, they unanimously turned him down and expressed their desire to keep him as pastor of Le Chambon. But the head of the church, eager to pull this thorn from its side, wrote Trocmé a letter accepting his resignation. On October 16, 1939, André Trocmé wrote to the head of his church, telling him that he had *offered* his resignation to the presbyterial council but the council had refused it. In that letter he said that he had offered that resignation for the sake of "the peace and the honor of the church," which could ill afford to have such conflicts within it. But their firm refusal to accept his resignation, Trocmé went on, made him pastor of Le Chambon "more than ever." He would continue to defend Christian pacifism and the religious liberty of all members of the church.

André Trocmé had been a rebel for years within the Reformed Church of France. The conversation his son had overheard was only one episode in that rebellion. Jacques's refusal to go to temple after overhearing that conversation in 1943 was a young man's version of his father's own refusal to compromise with authority when it came to matters of life and death.

2.

Jacques and Nelly are the only surviving children of André and Magda Trocmé, and when I asked them whether they were ever displeased with the refugees for taking food out of their mouths, they both gave the same answer: "Never. And you know, in the thirty-five years that have passed since then, it has never crossed my mind to think about this until this very moment when you raised the question."

And the food they ate was sparse: two lumps of sugar a day per child, beans that were not only tasteless but incapable of being softened by cooking, so that they made a *ping* on the plate after having been cooked for hours. Jacques remembers that he did feel some resentment about food—resentment against his sister and brothers. Each of them received one thick, triangular slice of bread a day. They watched the distribution of bread in the morning with an unswerving eye, and each yearned to get more than the others, or at least not less. And sometimes they stole bread from each other.

In the intimacy of a household, people come to understand each other in ways that their public lives can never reveal. Within a family, in a kitchen, in a dining room, the public masks are off. In the presbytery, Nelly saw her father from up close as a totally upright man, capable of towering rage, but swiftly repenting that rage, and unswervingly generous to strangers. She saw a man who would not drink alcohol or smoke tobacco, a man who saw dancing as a terrible temptation to sexual license. She saw a man she held in awe, a man with whom she had occasional communication and momentary hugs and kisses (all his life, even into her middle age, it was easy and natural for her to cuddle up in his lap). Always she felt that his sense of duty almost exuded from his pores. He was a man who was almost always exerting his self-control.

At home he would never express his morality in terms of commandments laid down by some *imperator,* some general who was dictating orders to inferiors. He expressed his morality by embodying it. He never preached to the children, never forbade them to drink or to smoke or to dance; he simply—somehow—made it impossible for them to do these things, given the father they had. He was warm and loving with them, but his daughter remembers how the moral pressures that were always at work within him were sometimes, to use her words, "hard to take" for the children. Affectionate as he was, he embodied a morality too strict for his energetic children to follow faithfully.

Their mother was different. She worked no less hard than André, and she was as dedicated to helping the refugees as he was—she worked even more passionately at her tasks than he did. But she was *lenient* with her children. She looked into their eyes and saw needs her husband could not see, needs that she allowed them to satisfy even though doing so sometimes caused pain to her husband. In their intimate moments with her, they never saw her as "upright" or "hard to take." She was far more supple under their wills than their father was, and far more aware of their feelings.

Once Jean-Pierre, the gifted child who would die in the last months of the Occupation, said to his mother, "*Maman,* I love you very much. I love you so much that I think I cannot get married because I cannot leave you. But if I do get married, why, I shall live in Tence [a few miles away]. And—I'll tell you what, *Maman*—when I have my first child, I shall give it to you."

On another occasion he said, "You know, *Maman,* Father is really a good man. When he dies, I am going to have a wonderful funeral for him—you know? And he will be in a wonderful coffin all covered with precious stones. And I am going to have made a mechanical bird, which will stand upon his tombstone, and this bird will say again and again, 'Here is the pastor of Le Chambon.' "

The differences between the two gifts to the two parents should not obscure the great love that both of those gifts expressed.

3.

There is in André Trocmé's papers a copy of a letter he wrote in February 1943, a few days before his arrest. It tells of Magda's losing weight, of her eating badly and aging swiftly, so that the family worried about her; and it tells of André's back troubles, which were making walking more and more difficult for him. And it tells of *surmenage,* too much strain on both of them, what with a house packed with people coming and going all hours of the day and evening. It tells also of the financial difficulties of the Trocmés in their efforts to care for themselves and the refugees with a small income and steeply rising prices that resulted from so many purchasers being in Le Chambon, and such limited supplies. And it succinctly summarizes the activities of the presbytery and of the village as a whole:

> . . . in the course of this summer we have been able to help about sixty Jewish refugees in our own house; we have hidden them, fed them, plucked them out of deportation groups, and often we have taken them to a safe country. You can imagine what struggles—with the authorities—what real dangers this means for us: threats of arrest, submitting to long interrogations. Today, because the Germans are now occupying the Free Zone, we are closer to the French authorities here . . . these facts about the help being given to refugees in Le Chambon are advertised throughout the south of France: from Nice to Toulouse, from Pau to Macon, from Lyon to Périgueux, passing through Saint-Étienne, it is by tens, by *hundreds* that Jews are being sent to Le Chambon. My daily ministry

> has been completely halted. I used to turn our dining room into a waiting room during the summer. Now it is a waiting room all year round.

The word that best summarizes the life of the Trocmé family in the middle of the Occupation (and indeed from this time through the end of the Occupation) is not "glory," and it is not even "tragedy." The best word is *surmenage,* the strain of too much work, too much tension—simply too much. There was danger, but the danger was quotidian, part of people's ordinary life; it was not the kind of danger that soldiers and their leaders experience in crucial moments of battle.

When toward the end of 1942 the Germans moved into the south, there were no occupying troops in the little village of Le Chambon. At the Hôtel du Lignon there were a few wounded or fatigued German soldiers, but when they appeared on the streets they were always correct, even helpful. Once Jean-Pierre was gathering pinecones for kindling, and a strolling convalescent German soldier helped him fill his basket and then quietly went away.

True, from time to time as they prepared to go to sleep, an urgent telephone message would arrive. Over the phone one would hear gruff, rapidly uttered words like "Attention! Attention! Tomorrow! Tomorrow!" and then a loud click as the instrument on the other end was swiftly and solidly hung up. One would tell a few people that there was to be a *rafle* (roundup) or a check of papers, and then go to bed. But this happened seldom. Usually, people moved without fanfare in and out of one's life. True, the stories of the refugees were often horrific, and the presbytery, like the town, was always conscious that it was "under the German cup," as a Chambonnais expression put it, but there was usually no military or dramatic "glory" or "tragedy" in it all. There was mainly coping—and work.

The *surmenage* fell on nobody in the presbytery more heavily than on Magda Trocmé, especially with the lack of money and the rising prices and increasing numbers of hungry refugees. Besides, as Trocmé put it in his autobiographical notes, Magda "invited many to eat with us, was in fact always inviting people." She was a part-time teacher of Italian in the school Trocmé and Theis had founded; and she worked with various groups in the parish, including a much-needed sewing group whose gossip almost drove her mad. Interminably involved with refugees and their special, often terrible problems, she also managed the big house, her energetic children, and her ever-inventive husband, though "managing" him would have been a full-time job in itself in a normal woman's life.

One year younger than the century, she had come to Le Chambon in the prime of life, and had grown stronger and more beautiful in the rudely beautiful and demanding climate of the village. When she married André Trocmé, she was a young, slender, and sickly woman, but over the years she became, to use Trocmé's own words, "classic, beautiful, strong, and ardent." By 1940, her beauty had become severe, Florentine—having nothing to do with delicate features, and having everything to do with unending energy. Her powerful indignation against laziness or ineptitude, for example, struck fear into husband, children, or guests whenever it was aroused. Her speedy, jerky, Italian-accented speech could fill the presbytery with her suddenly aroused and suddenly departing feelings. She seemed to thrive on inquietude. She was perpetually anxious over details, perpetually agitated over getting the details *just* right.

In the course of the growing problems of the Occupation, this agitation almost killed her. Many a day she would not feel like eating because she was so *surmenée* (tired and strained). And so she lost much weight and aged terribly. Sometimes all she could feel was weary disgust over the endlessness of the work. Busy as his life was, her husband had his moments of solitude and contemplation. He had his moments with his "musical rats." But for

Magda the rats—who once destroyed all of the children's bathing suits—were eaters of clothing, and they left behind them filth that had to be cleaned up. She remembers that she had to let the droppings harden a little because if she wiped them when they were fresh, they smeared things up terribly.

4.

Then there came into the presbytery a person who would keep Magda from dying of overwork and would give calm support and love to the children. In the winter of 1942–1943, Magda knew that she needed help, and needed it desperately, but when a woman visiting Le Chambon from Avignon told her that there was a strict Protestant community near Avignon that would give her the services of one of its female members, Magda became frightened. What would a pious woman feel about such a worldly creature as herself? She knew that people in the Avignon community prayed three or four times a day, and she also knew that except for the brief thanks to God that the family sang before every meal except breakfast, she had no time to pray in the presbytery. And all that Bible reading those religious ones did! All those meditations! She wondered, "What will the religious one think of me, a pastor's wife, running around and not stopping for God?" And so, sick with fatigue as she was, she kept postponing the date in January 1943 when the redoubtable woman should come. Finally she could postpone the date no more; she was too weak even to fear this new religious conscience that would be entering into the center of her life.

And so one day Alice Reynier appeared in the kitchen, with her hat askew above her round face and with a happy "Hello, friends!" She, too, had been afraid; André Trocmé was one of the most respected religious leaders in the south of France, and she was only a prayerful, bookish, tiny woman, who, when Magda's call came, had dropped to her knees and prayed to God that she might help the family who needed her. She was terrified at the prospect of being involved with such a great force as André Trocmé.

But she need not have been afraid. Within a short while she became the balance wheel of the presbytery, an integral part of the family, who would help keep them all from pounding themselves and each other to death with all their ever-active energies. Even though during her stay in the presbytery André Trocmé would be arrested and sent to an internment camp, her darling Jean-Pierre Trocmé would meet with tragedy, and the pastor would spend the last year of the Occupation running from the Gestapo, she managed to keep her commitment, one that she had made on her knees before she came to Le Chambon, to serve the family with joy, in peace, and in love. Early in her stay she asked everyone to call her "Jispa." The name, which she invented by taking a few letters from the French phrase that means "the joy of serving in peace and in love," was important to her. She had a very quick temper—her round brown eyes could flash suddenly into anger, and her anger took longer to subside than did André Trocmé's. She chose to be called by the name Jispa so that every time someone addressed her, she would be reminded not to get impatient or angry, but to serve that person in joy, peace, and love.

Within weeks of her arrival, she was using her skill at calligraphy to forge the signatures on the refugees' identity and ration cards, and she was in total charge—with her orderly mind—of handling the money in the presbytery. None of the Trocmés had much concern for money, except when the box in which they kept all their ready cash was empty, and then only Magda reacted, with absentminded surprise. Jispa took over the box, and everybody had to ask her for money. When they had to be careful—which was almost always during the Occupation—she alone had the last word on whether to spend and when to save.

Jispa is not quite five feet tall, and when she is not angry, there is a peaceful yet scintillating and surprised happiness in her that leaps out of her blue eyes and strikes one deeply. Always, as if only an inch or so *behind* those eyes, her mind seems to hover, detached, as if not quite all her joy comes from seeing the things and people around her. That detachment is in part the poise of a subtle, highly educated Frenchwoman who despises what she calls "baths of sentiment," but it is also the detachment in eyes that are praying to God while they are open and while they are looking at the world.

Jispa came to the Trocmés from a Protestant religious community to help them for a period of three months. She has now been with them for more than a third of a century. Once, long after the war, when Jispa said that God had directed her to the presbytery, Magda burst out in that rapid, jerky, Italian-accented voice of hers with, "No, Jispa! If God directs you to do good, then He must direct you to do evil too! And if He does not direct you to do evil, then there must be a Devil, and that is *much* too complicated—believing in God *and* a Devil! No, Jispa, no!"

In answer, all Jispa did was smile happily at the big woman. For Jispa, all the complexities of theology and all the complexities of life in the presbytery in those hard, dangerous years dissolve in a simple love for God that Magda could never know. During the last years of the Occupation, that always helpful spirit of hers often helped save the Trocmés, and especially Magda, from weary despair.

5.

Long after the war, Magda Trocmé summarized for me what all the work meant for her at the time:

> I have a kind of principle. I am not a good Christian at all, but I have things that I really believe in. First of all, I believe and believed in André Trocmé; I was faithful to his projects and to him personally, and I understood him very well. Second principle: I try not to hunt around to find things to do. I do not hunt around to

> find people to help. But I never close my door, never refuse to help somebody who comes to me and asks for something. This I think is my kind of religion. You see, it is a way of handling myself. When things happen, not things that I plan, but things sent by God or by chance, when people come to my door, I feel responsible. During André's life during the war many many people came, and my life was therefore complicated.

Her "principle" did not involve any abstract theories, but only a feeling of responsibility to particular people—first of all to her husband, and next to anybody who happened to come to the door of the presbytery. And this feeling is not one of overflowing affection; it is practical and abrupt, like Magda herself. The sentimental, sometimes mystical love cherished by poets, philosophers, and saints was not for her. In Plato's *Symposium,* Socrates tells us that love is born of the union of the lover's poverty and the beloved's plenty: it is a passionate yearning for completion. Her caring for those around her was quite different: she had much to give those in want, and felt a responsibility to give it—not with feelings, but with action. Hers was a caring involving help, not romantic yearning.

In Deuteronomy, a city of refuge is a place that takes responsibility for the life of the refugee who comes there; that is to say, it is a place that commits itself to protecting the refugee who comes to its gates. Its members do not leave those gates to look for the oppressed; they stand at the gates ready to take the responsibility people put upon them by coming to their city. Deuteronomy 19:10 reads: "I command you this day to [protect the refugee] lest innocent blood be shed in your land . . . and so the guilt of bloodshed be upon you." Magda Trocmé hardly ever speaks of God, and she does not think in terms of innocence or guilt; rather she thinks in terms of people in trouble, in terms of doors that open, and in terms of her own stubborn decision to keep her door open no matter how tired or how distraught she may be. Despite her secularism, she was an effective gatekeeper for a city of refuge.

In fact the word *love,* with its usual connotations of affectionate yearning, does not occur much in her speech. She believes love involves preferring somebody to somebody else, the way she preferred André Trocmé to all other men. In this sense of the word, she did not love the refugees who came through the "poetic gate" on the Rue de la Grande Fontaine and appeared in the shallow doorway of the presbytery. Her husband acted out of a passionate, even a mystical love, strongly tinged with the supernatural; but not she. Her religion of love was summarized in an abrupt, ungrudging, raucous command issued through a wide-open door: "Naturally, come in, and come in."

Not only is she reluctant to use the word *love* when talking about her work with the refugees; she is also reluctant to use words like *good* and *saintly.* She does not believe that there is such a thing as moral nobility that sets off some people—the saints—from others—the common, decent people. To use a language she profoundly distrusts—the language of theology—she does not believe in works of supererogation, deeds of extraordinarily high ethical value. In this belief in moral democracy she is—unwittingly—very close indeed to the beliefs of most Protestants: there are no Protestant saints; there is no Protestant moral nobility. There are only people who accept responsibility, and those who do not. For her, as for them, a person either opens the door or closes it in the face of a victim.

But though there is no moral nobility for her, it is still far better, according to her practical "religion," to open the door than to keep it closed. And the reason that it is better is simple for her: that person there—that pursued, terrified person before her—needs help, not a closed door. *That person's need* is the basis of ethics, not any sentimental love she may or may not feel for the refugee, and not any calling she may feel to be morally noble.

When I pushed her to account for this sense of responsibility, her impatience soon began to show itself. Once she lowered her head under such probing, looked away from my eyes as if she were not going to be completely straightforward with me (but this was all I deserved to get with such silly questions), and said, "I get pleasure from doing such things—yes, pleasure, the way some people get pleasure from the movies. It amuses me to help somebody, no matter what the cost." At another time she said, with the same lack of candor and conviction, "Well, who knows? I might need to be helped by them someday!" We had reached bedrock in her thinking; there was no way to go deeper; the spade had turned. She was losing interest in my silly questions because she was the answer to them, not her words or even her thoughts, only she, Magda Trocmé.

Her sense of responsibility did not put her in a condition of slavish obedience to every whim of every refugee who came to Le Chambon. Far from it. Once a friend of Magda found a little house for a Jewish husband and wife who were refugees. Magda felt quite free to open other doors than her own to refugees; in fact the presbytery, like some of the other houses in central Le Chambon, served mainly as a way station to other homes in the area, usually farms around the periphery of Le Chambon. When she greeted the couple at the railroad station near the post office, it was raining hard. In the station she gave them a map showing them how to get to their new home. The woman complained, "But oh! This is far! How can we get there? And it is raining so hard!"

For this she received the full weight of Magda Trocmé's Italian and Russian passion: "A few drops of rain are bothering you? What of my friend Simone Mairesse, who found the house for you? The rain fell on her. She ran around at night, put herself in danger, and she is a widow, and she has a child, and she could have been killed, and you are saying that you can't walk this distance because of the rain?"

The woman was silent, bewildered by the downpour within a downpour. Such events, incidentally, were rare in Le Chambon; there was usually a confident meeting of the refugee with the Chambonnais, and the refugee would usually reach out to the Chambonnais with affection.

Nelly called her mother a "tormented Italian," possessed of immense vigor and practicality, and married to a man she understood deeply and lovingly. In the presence of her mother she once said to me, "If she had married a businessman, she would have turned her energy into that. Her dedication was not because of religion; it was because of people."

But she and her husband cherished the speedy thinking and rising flood of passion they always found in each other. They were passionately in love with each other from the day of their wedding to the day of his death, and during the years of the Occupation they agreed with each other completely about what was the right thing to do in Le Chambon.

Still, they were two very different people, and they often had heroic battles. They argued loudly about theology, politics, and the raising of children. He was a committed, deeply devout Christian; she was suspicious of Christianity and every other religion. He was a liberal, though raised as a member of the bourgeoisie; she was often described as *communisante* (Communistic), or at least edging toward Communism in her desire to wipe out the whole class system that her husband cherished in his heart. He was strict with the children; she was very lenient with them.

One day at dinner at the big dining-room table, they found themselves in stubborn disagreement with each other. André was talking vigorously, but Magda had her head down and was disagreeing unshakably with him. He had a glass with a little water in it near his hand; suddenly, he took up the glass and threw the water at her bent head. She said nothing, took up a pitcher of water as if she were going to pour herself a drink, and then stood up and threw the whole pitcherful straight into his face. Nelly was sitting on her father's right at the time, and as she watched her mother pick up the pitcher, she found herself thinking, This would be fantastic if she really did it. Right back and ten times more. Will she or won't she? And when Magda did it, the whole table—parents, children, guests—all exploded in laughter. Magda herself laughed until she cried—she always cries when she laughs hard. It is no wonder that a Chambonnais once described their arrival in Le Chambon as the "arrival of Vesuvius and Stromboli."

They argued often, but usually their arguments happened during the midday meal, when there was a bit of time for a talk (he usually had meetings immediately after the evening meal). And so they had time to kiss each other and to understand and forgive each other before falling asleep. On only one night in their long marriage did they fail to do this—and the fact that the argument took place at night, before a meeting, was the only reason for the failure. That night Magda decided that she would restrain her impulse to take him in her arms, so that he would realize that the argument was not entirely negligible. But he was tired after the meeting, and they both fell asleep. More than a third of a century later, Magda remembers that night, not with regret—she has few regrets—but with mild amazement that this could have happened, this neglect between two people who loved each other so.

One of the main differences between them was that André Trocmé was profoundly innovative, creative; he was a spiritual and theoretical volcano, always producing new ways of conceiving things. Magda was preoccupied with coping. And so was Trocmé's other partner, Édouard Theis. When the war was over, Magda and André made an agreement that he would stop inventing new plans, at least for a while. A few days after the agreement, he laid a half-dozen fully incubated projects upon the dining-room table—he was addicted to creativity. Neither of the two main partners he had in Le Chambon could match, or would even try to match, his creative imagination and his desire to change things fundamentally, but their passions for coping complemented his power to dream.

Unlike Magda, he was the kind of person who would *not* stand behind a door, ready to help, ready to take responsibilities thrust upon him by the outside world. He pushed his big body and his imagination out of doors, and he pulled people with him with an aggressive love. The word *aggressive* has roots in Latin words for stepping *forward,* as well as for acting *first.* It is a word that implies being at the forefront both in space and in time; it implies being on the attack. André Trocmé's manner was aggressive in these senses of the word.

As one refugee put it, he gave Le Chambon his personality; at least as far as attracting refugees was concerned, this individual went on, he impressed his personality upon Le Chambon, so that the village, like him, drew people like Magda and Édouard Theis to organize the details of their living so that they could be housed and fed, sometimes for years, in the poor village and on the poor farms. Various Chambonnais called him *un inspiré,* a person inspired by his devotion to Jesus to move and to shake the lives of those around him.

Another refugee replied, when I asked her what kind of man he was: "That smile . . . that smile . . . the smile of that man, that smile . . . He did not have to say anything, just *'Ça va?'* and that was enough." Still another said that upon arriving at the presbytery she was greeted with immense warmth, as if he would fold her in his arms and protect her lovingly against any harm. A few minutes after she met him, he offered to lend her money, which she did not need. A little later, she found out that he had almost no money himself.

One of the refugees was Daniel Isaac, son of the distinguished French historian Jules Isaac. Being by definition a Jew because all four of his grandparents had been practicing Jews, he found himself living amid the growing anti-Semitism of the French people, though he was a thoroughly assimilated Frenchman and a Protestant. The radios, the newspapers, the films, and even wall posters were making Frenchmen more and more anti-Semitic in accordance with Vichy policy. One day he received a phone call from Pastor Trocmé of Le Chambon. Trocmé asked him to come to the village to teach philosophy to summer school classes at the Cévenol School. Isaac replied that he had no experience in teaching philosophy and had studied very little of the subject. Besides, he could not move his family so far for the sake of a temporary summer job. Trocmé insisted, telling Isaac that he would be rendering the school and Le Chambon a great service if he came. Isaac and his family had spent some time in Le Chambon; they were all practicing Protestants, and Le Chambon was the Geneva of France as far as Protestants were concerned, and so he felt a sense of obligation to the pastor. He went, and in the end left his family there for the rest of the Occupation while he went elsewhere to work for the Resistance.

He is certain that André Trocmé was trying to protect him as a Jew. It was characteristic of him, Isaac says, to help you while convincing you that you were helping *him.* In fact, many of the refugees who passed through Le Chambon found themselves receiving André Trocmé's gratitude while he was saving their lives.

And this was not a device; it was not a polite, condescending form of cunning on Trocmé's part. He did need to have the refugees there; he did need to protect them; and he did need their help in some project or other that his fertile brain had just created.

Unlike Magda, he was capable of a love that was affectionate without being preferential: any human being, simply by virtue of being a human being, was profoundly valuable to him. He could grow mightily angry at people, and he could criticize them very painfully sometimes, but nonetheless he could embrace them with his affection immediately afterward.

Only he—not Magda and not Édouard Theis—could have gone to Marseilles to talk with Chalmers about what to do beyond the doors of Le Chambon itself. But though he was aggressive, he did not smother the objects of his affectionate love. On the contrary, he stimulated them. Daniel Isaac has compared his presence in the presbytery to a glorious performance of Beethoven's *Eroica:* it lifted you, excited you, warmed you. It made you rise to your own highest level of joy and vitality. And it did this not by command but by contagion.

Magda felt a simple responsibility to help a person whom "God or chance" had brought to her door. Trocmé had a religious conviction that gave shape and direction to his actions as much as his own warm temperament did. Throughout his mature life, he stated that conviction in many forms—in conversations, in sermons, in lectures, and in writings—but he never stated it more concisely than he did in an article in 1955:

> Basic truth has been taught to us by Jesus Christ. What is it? The person of any one man is so important in the eyes of God, so central to the whole of His creation, that the unique, perfect being, Jesus, (a) sacrificed his earthly life for that one man in the street, and (b) sacrificed his perfection [by taking the blame for his sins] in order to save that single man. Salvation has been accomplished without any regard to the moral value of the saved man.[6]

Trocmé believed that beneath and pervading the personal qualities, the moral strengths and weaknesses of a human being, there is something immensely valuable, something beyond all price. If that human being is saved by God, or if that human being is saved from harm at the hands of other human beings, he is saved not because he is good; he is saved in both cases because he is a being whose importance God has shown us by sending Jesus to help us. And because of this valuable something that is beneath and pervades our personal qualities, a person is not to be killed because he is wrong or because he is evil. For Trocmé, every person—Jew and non-Jew, German and non-German—had a spiritual diamond at the center of his vitality, a hard, clear, pricelessly valuable source that God cherishes.

This basic truth was what Trocmé was expressing when, on the day Georges Lamirand, the Vichy minister in charge of youth, came to Le Chambon, he said to Prefect Bach, who had just reprimanded him for not seeing that the Jews were corrupting the West and must be gotten rid of: "We do not know what a Jew is. We know only men." It is this belief that he communicated to the Chambonnais who let the Vichy police lieutenant fall into a cesspool in his search for Jews. After they had cleaned him up and comforted him, he asked them where the Jewish refugees were, and they replied, "Jews? What would Jews be doing here? You, there, have you seen any Jews? They say they have a hooked nose." The people of Le Chambon, like their pastor, saw only human beings who were valuable enough to be saved from humiliation, torture, and death.

6.

Magda Trocmé believes that something is evil *because* it hurts people. Hers is an ethic of benevolence: she needed only to look into the eyes of a refugee in order to find her duty.

But her husband had a more complex ethic. He believed that something is evil both because it hurts somebody *and* because it violates an imperative, a commandment given us by God in the Bible and in our particular hearts. He had to look up to some authority beyond the eyes of the refugee to find that commandment, but having found it, his duty, like hers, lay in diminishing the hurt in those eyes.

Magda's ethic can be called a horizontal one: she recognized no imperatives from above; she saw only another's need, and felt only a need to satisfy that need as best she could. She would work for those who needed her help—and she almost worked herself into an early grave for them—but she did not work out of awe for any superior being beyond mankind.

Her husband's ethic drew its power from the life and death of Jesus. The example and the words of Jesus inspired awe in André Trocmé, and he did what he did because he wanted to be *with* Jesus, in the sense of imitating Jesus's example and obeying his words. His obedience to Jesus was not like the obedience of a soldier to a military leader; it was more like the obedience of a lover to his beloved. He wanted to be close to Jesus, a loving disciple who put his feet in Jesus's footprints with stubborn devotion.

There was verticality in his ethic, an allegiance to a supernatural being, but there were also in him powerful affections, "almost erotic" feelings for the people around him. In his sermons, especially during the Occupation, he dwelt as much upon practical, worldly plans for resistance and decency as he did upon the life and words of Jesus. When people left the temple after those sermons, they often said, "Ah! So *that* is what I should do about it!" He worked and cared for the well-being of the "oppressed and the weak," as he once described the refugees, as much as did Magda Trocmé, but he never stopped striving to be close to Jesus and, in Jesus, to God. For him, ethical demands had a vertical axis and a horizontal one, like the cross.

But Magda had long before saved him from a mystical confinement to the vertical axis. For instance, nonviolence, which was always close to the center of his concerns during the Occupation, was not simply a belief held by an obedient disciple of Jesus; it was practical for Le Chambon then, during those years, and it was the most practical course of action available to the village. Only nonviolence could allow Le Chambon to resist the will of Vichy and the will of the Germans without a massacre. The little village was surrounded by soldiers ready and, in some cases, especially after the Germans moved down to occupy southern France, eager to do violence to anybody who tried to resist them with violence. But nonviolence was disarming, and in the face of the overwhelming power that surrounded Le Chambon during most of the first four years of the 1940s, it was a course of action dictated by practical benevolence toward one's fellow human beings.

It was because of this shared commitment to worldly decency that Magda Trocmé could say, after the waterfight, after all those arguments about theology, politics, and children, and after all those times when her husband pursued his own aggressive calling beyond the doors of the presbytery (doors that Magda would have been content to stand behind), "André and I were in agreement." They were together in one ethical dimension—caring for others. And because they were so much at one with each other in that dimension, their differences usually seemed funny, minor, and fleeting. The open door and the "Naturally, come in, and come in" made the presbytery a community.

7.

But it was a community totally different from the isolated community of the "Trocmé people" in André's birthplace, Saint-Quentin. Both the "poetic gate" and the shallow main entrance to the presbytery itself were almost always unlocked. People streamed in and out of the house without knocking if they were Chambonnais, and sometimes if they were not. Some refugee German girls once told Magda (with that special trepidation that German refugees felt) that she should lock one of the doors. Magda would not do it; for her, there could be no hard line between the presbytery and the rest of Le Chambon. A few days after the conversation with the refugees, Magda returned from working all day at the Cévenol School and found flowers everywhere in the house: in the kitchen, in the dining room, in her husband's office. She still does not know who put them there—it might have been the refugee girls. But never mind, she thought, let flowers come in as well as horror.

Through these open doors, people would bring food (Simone Mairesse, Magda's dear friend who had found the house for the refugees who arrived in the rain and complained, regularly brought a large part of a pig, and Magda would then have to apologize to her Jewish guests for giving them pork). People bringing and asking many things came through those doors, like veins and arteries carrying blood to and from a heart, and tied the household not only to Le Chambon but to Europe and America.

There is one story that can convey the relationship of the presbytery with the outer world as well as any. One evening a woman of the parish swept through the open doors, rushed through the dining room, and pounded on the outer door of Trocmé's office.

Once inside the office, she started to implore her pastor to officiate at the funeral of her husband. Trocmé knew that she had shamed and badgered her husband into giving up nonviolence and joining the Maquis. Now he lay dead in a field between Le Chambon and Le Puy. Would her pastor, who had been preaching nonviolence to them both for almost a decade, officiate at the funeral of her Maquisard husband?

In the last two years of the Occupation, in such circumstances, it was a crime punishable by death to preach any service over the body of a dead Maquisard. But Trocmé did not hesitate. Getting precious gasoline from a black marketeer everybody called the "gangster," he drove out of Le Chambon to perform the ceremony. When he arrived at the place where the Maquisard had been killed, he found no corpse. He learned from peasants who lived near the place of the shooting that a priest had performed a Catholic service over the body and buried the Maquisard in a nearby village cemetery. Trocmé had the corpse removed to Le Chambon's churchyard, and it is now buried next to the ashes of André Trocmé himself.

We fail to understand not only the role of the presbytery in Le Chambon but also the nature of the Resistance there if we think that there was a sharp division between advocates of violence and those of nonviolence. Friendship and a refusal to surrender their consciences to the Nazis and their collaborators were far more important to the Chambonnais than any issues such as violence or nonviolence. More, the image of a Catholic priest praying at the risk of his life over the body of a dead Maquisard without knowing whether the man had been a Protestant or a Catholic is a far more accurate representation of the Resistance than any discussion of the differences between Catholics and Protestants would be.

When in August 1944 the Trocmés' eldest son, Jean-Pierre, died and had to be carried to the cemetery for burial, among the pallbearers were members of the Maquis. They were not there as representatives of a group; they were friends and allies of the

Trocmés, whose lives were intertwined with each other as loving friends and as people taking a stand together against what they were all convinced was evil.

But that funeral must not be taken to be a symbol of the presbytery and its function in Le Chambon, just as its thick walls must not be taken to be a symbol of that function. Life was what the presbytery meant in Le Chambon, vivacious, generous life. The Trocmés in the presbytery brought this to Le Chambon. Life—for the weak and for the strong—was their cause, and at great cost to themselves and to others, that cause triumphed in Le Chambon. After all, somebody brought them flowers through those open doors.

7

The Inspired Amateurs

1.

There is a French word that applies to the Trocmés and to the spirit they created in the presbytery: the word is *accueillant.* English words like "welcoming" and "receptive" convey some of its meaning. It was because the Trocmés were by temperament and by conviction *accueillants* that the idea of a place of refuge could become a reality in Le Chambon. If their abundant energies had been confined to preserving the peace of mind and the comfort of themselves and of the parish that was in their charge, the idea Trocmé brought back from the Nimes conversations with Chalmers would have remained only an idea. But in the openness of the presbytery, the idea found a fructifying atmosphere and flourished.

When in the terrible winter of 1941 Trocmé returned from the last of the Nimes conversations with Chalmers, he found a wife as ready as he was to make Le Chambon a city of refuge. But without a presbyterial council that was itself *accueillant,* and without a parish that supported that council, the story of Le Chambon would not be the story of a city of refuge. As soon as he returned from his last conversation with Chalmers, Trocmé convened the council and presented to them the idea of opening their village and their own houses to strangers who would bring danger with them. Without going into details, he tells us in his autobiographical notes that he "won an easy victory" and the council committed the parish to the task in one meeting.

One of the reasons for the swiftness of their action was the personality of Trocmé, and another was the solidity and power of the ideas he and Chalmers had discussed in Nimes. But there is a deeper reason for their speedy decision. The history of the Protestants in France, and specifically the history of the Protestants in The Mountain, in the two adjoining villages of Le Mazet and Le Chambon, had prepared them for a certain kind of resistance to governmental authorities. Farther south in France, the revocation of the tolerant Edict of Nantes in 1685 had produced, in time, the bloody battles between the government and the Camisards, the Huguenots deep in the south of France. But during these battles, which took place during the first decade of the eighteenth century, Le Chambon with its sister village Le Mazet engaged in a different kind of resistance. They fought no bloody battles with dragoons, but instead used the devices peculiar to mountain people: silence, cunning, and secrecy. They resisted the authority of the government as firmly as did the Camisards, but they resisted by quietly refusing to abjure their faith, and by quietly conducting their services in meadows within the pine forests, and with a portable *chaire* (pulpit). This was the kind of resistance peculiar to The Mountain: the resistance of exile.

Such a tradition fit beautifully with the ideas Trocmé and Chalmers had formed in Nimes. The Quaker and the Huguenot agreed that, to use Chalmers's words uttered long after the conversations, "there was no limit to what might be possible in terms of the reclamation of persons." They did not want to kill any follower of Pétain or Hitler for any reason, because that human being *could* be "reclaimed." And they believed that the victims of Vichy and the Nazis were also reclaimable, could also be "saved." Trocmé himself, being a Huguenot, was more ready for clandestine activities than were the Quakers, who chose always to "speak truth to power," but now the differences between the two groups were unimportant; both were bent on resisting the violation of their own consciences and of other human lives not by killing but by saving.

After he won his "easy victory" with the presbyterial council, the first major step Trocmé took was to choose a head for the first house of refuge in Le Chambon. According to the agreement with Chalmers, this "monitor" would be in charge of feeding, clothing, protecting, and educating young children whose parents had been deported. He chose his second cousin Daniel Trocmé, a young teacher of languages, history, and geography, who was now in southwest France.

Daniel Trocmé would die for that conscience before the Occupation was over, but this was, of course, unknown to the pastor, and so he made Daniel the head of the first house, the Crickets, which was a dilapidated old boardinghouse for children situated about two miles away from the center of the village. When in the course of the Occupation the Quakers left southern France, they kept sending money faithfully to Daniel and the Crickets by way of Geneva. Sometimes the couriers who brought this money were arrested, and some were shot by the Germans, but the money kept coming.

2.

The development of Le Chambon as a place of refuge proceeded more and more swiftly as more and more Chambonnais and more and more organizations and governments outside France became aware of the need for such a refuge and of the capacity of the village to satisfy that need. A mid-Occupation letter gives a summary of the situation in Le Chambon shortly after the Germans moved down to occupy southern France: "3,300 parishioners, of whom 2,000 are peasants, 700 are villagers, and from 500 to 600 people from outside Le Chambon: 160 refugees from central Europe, adults, students, children in six [seven, including the Farm School] different houses; 300 students at Cévenol School; plus 30 teachers and 15 heads of [school] residences . . ." But there is another way of understanding the anatomy of the rescue activities of Le Chambon. Imagine a marksman's target with its concentric rings around a bull's-eye. Put the groups of Le Chambon that inspired and led the rescue of the refugees closest to the bull's-eye, and put the groups peripherally involved with those efforts farther from the center of the target. Outside the target, put the enemies of the refugees, Vichy and the Nazis.

Those who were most directly involved in the rescue effort were those who started and guided it, and so the presbytery should be at the bull's-eye of the target. Not only did André Trocmé bring the idea of a city of refuge to Le Chambon; he also set an example by keeping refugees in the presbytery throughout the Occupation. Moreover, many of the refugees who came to Le Chambon passed through the doors of the presbytery. After the arrival of the one o'clock train (or some other means of transportation), many would come to the presbytery, and would receive some food until a temporary house in the village itself could be found for them; then, if they were not staying at the presbytery, they would go to this temporary shelter. While they were there, false cards of identification and false ration cards would be made by Darcissac and others. Then, after a few days, equipped with their cards, they would usually move to a more permanent shelter, or, later in the Occupation when Le Chambon was overcrowded with refugees, they would be made part of a team to be conducted across the mountains to neutral Switzerland.

In this operation, the presbytery played an important role, but so did the temple, and so it should reside in the ring immediately outside the bull's-eye. From the *chaire* against the west wall of the temple, Theis and Trocmé delivered their Sunday sermons. Not all the parishioners fully approved of the worldly contents of those sermons, but they were the unfailing source of the major ideas and attitudes of the community.

Theis has said that his and Trocmé's favorite parts of the Bible were the Good Samaritan passage in Luke (10:27–37) and the Sermon on the Mount. Before the Good Samaritan passage, Jesus had been asked how one could achieve eternal life, and he had answered: "You shall love the Lord your God with all your heart, and with all your soul, and with all your strength; and with all your mind; and your neighbor as yourself."

But the questioner, who was learned in the law, pressed Jesus and asked, "Who is my neighbor?" Jesus answered with the Good Samaritan story of a humble person who "fell among robbers, who stripped him and beat him, and departed, leaving him half dead." One's neighbor is anybody who dearly needs help, their sermons said, anybody in terrible need of assistance. In speaking about these passages, the two ministers communicated to the village the main driving force behind the Resistance in Le Chambon.

Trocmé's sermons emphasized the need to obey one's own conscience when there is a conflict between it and the laws and commands of governments. He often talked about the "power of the spirit," which he described as being a surprising power, a force that no one can predict or control. He offered no systems or methods—this would be to violate the surprising force of the spirit—but he had one principle that he never forsook: the obligation to help the weak, though it meant disobedience to the strong. Apart from that repeated principle, he himself embodied the surprising force he spoke about so often. Nobody knew what their pastor would come up with next, so full was he of fresh responses to new situations.

During the Occupation, André's brother Francis wrote a letter describing his impressions of one of his sermons:

> . . . he is a pulpit orator who is absolutely original, who surpasses in authority anyone I have ever heard speak from the *chaire.* He begins in a simple, familiar mood, starting with recent events, everyday or religious, then he raises himself, little by little, analyzes his own feeling and thought, confesses his own heart with a sincerity and a perspicacity which disturb one; he uses the popular language, and sometimes crude language. . . . Is he not going to fall into trivialities? . . . But no! See him there raising himself up . . . he climbs, climbs always higher . . . he draws us to the peaks of religious thought . . . and once we are at the summit, he makes us hover in a true ecstasy; then gently . . . he descends slowly to earth and gathers you in a feeling of peace which gives the last word "Amen" all the meaning the word has etymologically. One sits there afterwards . . . eyes clouded with tears, as if one has been listening to music that has seized you by your entrails.

Part of the dossier supporting the award of the Medal of Righteousness to André Trocmé by the state of Israel is an account of the sermon he gave during the *rafle* of the empty buses described in Chapter 4. According to the state of Israel and the notes of André Trocmé, on August 16, 1942, he described the horrors of the *rafle* at the Paris Vélodrome d'Hiver that had occurred just one month before the sermon. Here are some of the words he uttered about that event and other events like it happening in Europe at the time: "It is humiliating to Europe that such things can happen, and that we the French cannot act against such barbaric deeds that come from a time we once believed was past. The Christian Church should drop to its knees and beg pardon of God for its present incapacity and cowardice."

A sermon in a Huguenot parish is not a performance that is supposed to be esthetically enjoyed by the auditors, and then ignored or praised the way one praises a fine performance of a piece of music. In this regard, Francis Trocmé's comparison of the effect of one of André's sermons to the effect of listening to music is misleading. Behind a Huguenot sermon is the history of a besieged minority trying to keep its moral and religious vitality against great adversity. The sermons of the pastor are one of the main sources of this vitality.

In Le Chambon in particular—a stronghold of Protestantism in France for more than four centuries—an effective pastor is especially precious because he keeps for his parish a little place for the love of God and of one's neighbor in the middle of a cruel, powerful outer world. He is not totally unrelated in their minds to Jesus Christ, who kept such a place for the Jews under the political and military power of the Roman Empire. And the pastor-martyrs of Le Chambon lend a special poignancy to his presence and to his sermons. For instance, Laurent Chazot, the pastor of Le Chambon in the early years of the Protestant Reformation in France, was burned alive for his preaching in August 1529. For Huguenots whose ancestors worshiped in this place near the Lignon River when there was no temple here because all Protestant places of worship were razed and forbidden and because all pastors were outlaws, a powerful pastor is indeed someone who feeds his hungry sheep, as the Latin word *pastor* implies. In their minds he is deeply involved with the idea of a Christ who sacrificed himself to feed the weak and sickly spirits of mankind.

It was not only the sermons that stimulated the rescue efforts in Le Chambon. It was also the *responsables.* During the Occupation, Trocmé replaced the dreary, perfunctory Sunday Bible readings with classes that put him in close, regular touch with the whole parish of Le Chambon. Every two weeks he met with thirteen people, and they all discussed a passage they had been thinking about for those two weeks. Trocmé did not lay down an interpretation and ask them to carry it away with them; he helped stimulate their own interpretations and let them flower.

Those thirteen leaders then went to thirteen different parts of the parish, and each tried to do with the people in his area what Trocmé had tried to do with them: stimulate a *valuable* understanding of the chosen passage. Of course, they had to arouse those responses in people not as enthusiastic or learned as the leaders themselves, but they succeeded admirably in the weekly sessions with their respective regions of Le Chambon. The local discussion groups doubled in size after a short time and attracted younger and younger people—always a healthy sign in parish discussions. Above all, the ideas that came from each of these groups were, to use Trocmé's words, "fervent, practical, and concrete."

The thirteen local leaders were called the *responsables,* and it was they who became the backbone of the parish as far as sheltering and hiding refugees was concerned. Through the biweekly meetings in the temple with Trocmé and the weekly meetings with each of the *responsables,* nonviolent resistance in Le Chambon developed its basic theory and its practical applications. For these were not sessions of sheer "enthusiasm" or mystical piety; more and more as the Occupation went on, practical plans for "overcoming evil with good" came from the *responsables* and the parish. In those meetings they all discovered what the Good Samaritan passage meant (who the neighbor was whom we were being told to love and to help) and they discovered the meaning of the Sermon on the Mount—not in terms of some abstract, pious theory, or even in terms of a long-term plan, but in terms of day-to-day decisions appropriate to the circumstances of Le Chambon. Here is the way Trocmé summarizes the work of the *responsables* in his notes: "It was there, not elsewhere, that we received from God solutions to complex problems, problems we had to solve in order to shelter and to hide the Jews. . . . Nonviolence was not a theory superimposed upon reality; it was an itinerary that we explored day after day in communal prayer and in obedience to the commands of the Spirit."

The *responsables* became the nervous system of Le Chambon.

Once, in conversation with Ernest Chazot, I asked how he would describe Trocmé as far as the parish was concerned. Without hesitation he said, "He was the great *responsable.* Yes, the great *responsable.*"

But this did not mean that the parish was smoothly harmonious through those four years. Some parishioners complained that the pastor spent too much time with foreigners and young people, and not enough time with the sick in the parish. And it was true that Trocmé was diminishing the number of his visits to people who were spending fifteen or twenty years in the process of dying. He could not be everywhere at once, but he felt that he was serving his parish through the *responsables.*

In fact, Le Chambon was far from being at peace with itself and with its pastor. One could hear all sorts of languages in the streets of the once isolated little village, and one could see there all sorts of people: Christians and Jews, Westerners and Orientals (from the Cévenol School), made this formerly homogeneous village a Babel of languages and ways of life. One of Trocmé's main problems was with the peasants in the outlying farms. They often felt themselves put in the shade, as far as political and economic influence was concerned, by the refugees, whom they sometimes confused with the rich tourists they were accustomed to seeing in the summer. And the refugees, as well as the students and teachers of the Cévenol School, were sometimes irritated by the slow pace of the thinking and action of the peasants and the townspeople. The youthful liberty of the students of the Cévenol School, and even of their faculty, as well as the sophisticated manners of many of the refugees, collided with the quiet, gray restraint of many of the native Chambonnais.

It was not a smoothly running community by any means. But a third of a century after the Occupation, a member of the faculty of the Cévenol School, Miss Maber, who teaches English there, mentioned the Trocmés to a peasant couple as they sat in their kitchen in an outlying farm. The couple wept with love and gratitude for the Trocmés.

3.

It is tempting to make the Trocmés all-important in the story of Le Chambon, but it is wrong to do so. Though the ideas and some of the energy behind a place of refuge came from them, very little would have been accomplished in Le Chambon without what we may broadly call "the houses" of the village. The presbytery and the temple could not have moved an inert mass of coolly self-centered people into doing what was done in Le Chambon. It was the houses, the homes in Le Chambon that made a village of refuge work. A family of refugees might come to town in winter, and the morning after their arrival they might find a wreath of holly leaning against their front door, with no hint of the identity of the giver. A little boy would come to Miss Maber's door, screaming in a high-pitched voice, so that the whole neighborhood could hear, that the English teacher had better hide Henri because the police were after him. Miss Maber would calm the boy, glance around to be sure that no strangers had heard him, and then go straight to the house of a mousy little Chambonnaise who was known to have an empty room. She would ask the tiny woman if she would hide Henri, and the woman would answer immediately, "Yes. There is a room downstairs, and the door opens into the woods. If the police come, he could have time to get away."

Though in my target diagram of the rescue activities of Le Chambon I put "the houses" outside of the bull's-eye and outside of the innermost ring of the target, this does not mean that the houses are less important to the rescue activities than the presbytery and the temple; it means only that the original idea and impulse for creating a village of refuge came from those places, and that those places guided the activities. *For the refugees themselves,* the houses were *the* scene and the source of those activities. What I have called the "kitchen struggle" of Le Chambon did occur in kitchens, in homes all over the village.

But there were different kinds of houses, and therefore somewhat different kinds of rescue activities occurring there. There were the funded houses, the places, like Daniel Trocmé's the Crickets, financed by great organizations in the outside world. The Quakers, American Congregationalists, even national governments like those of Sweden and Switzerland, helped finance those houses. By the middle of the Occupation there were seven of them.

One of these houses was the Farm School, funded by Switzerland. It was located far out in the countryside. Though its original purpose was to teach efficient farming methods to the students of the Cévenol School and to the farmers in the area of Le Chambon, early in its history it became a principal place of refuge, one of the safest homes in the area, not only because it had a clear view of all the roads from the nearby villages, but because the barking dogs of the surrounding farmers gave refugees plenty of time to disappear into the pine forests nearby.

Another funded house outside the village was the Flowery Hill. It was perhaps the most effective "underground railroad station" (to use the language of nineteenth-century American history) in the area. Just as American slaves in the nineteenth century stayed for a while in a place that protected them from capture and helped get them closer and closer to the North Star and freedom, so the Flowery Hill sheltered people who would soon be going to Switzerland. Besides certain Catholic groups, the World Council of Churches, and Sweden, the Flowery Hill was financed by one of the most remarkable rescue machines in the history of Europe, the redoubtable Cimade.

The Cimade was originated and led by women alone. In 1939, when they began, they helped care for displaced Alsatians who had been evacuated by the French from their homes on the border between France and Germany. But in 1940, after the defeat of France, they turned their attention to the most endangered human beings in Europe: Jews fleeing from Central and Eastern

Europe. At first they tried to relieve some of the suffering in the terrible, disease-ridden internment camps of southern France, and they made careful records of the horrors there in order to mobilize world opinion against them. But these activities did not satisfy their need to help, and so they developed a web of *équipes* (teams) in the summer of 1942, when the intentions of the Nazis toward the Jews were plain, and with these teams they took through the mountains of France to neutral Switzerland the refugees who were most dangerous to their hosts and most endangered themselves, the Jews.

In the last year of the Occupation, one of the members of such a team was Pastor Édouard Theis, who was then fleeing from the Gestapo. He joined them in leading groups of refugees through dangerous mountain terrain and through even more dangerous German troops to the Swiss border.

The Flowery Hill of Le Chambon was one of the houses in their complex network. With the thick woods to its north and a wide command of all approaching roads to its south, east, and west, its main function was to house the old, the sick, and women with small children only until the Cimade could make arrangements with Swiss authorities for their reception in Switzerland and could put together a team to take them to the border.

Aside from the funded houses, there were the *pensions,* the boardinghouses toward the center of the village. There was, for instance, the boardinghouse of Madame Eyraud, she of the round, pink face and the ready wit and even readier maternal affection. Her *pension* stood where two busy roads, the Street of the Soul's Song and the Street of Lambert, crossed. She kept only boys, usually fourteen of them, and they were usually not all Jews. Sometimes an adolescent would appear at her door with no baggage, because in Lyons or Saint-Étienne, where he came from, he had returned home from the factory and found the police surrounding his house waiting to take him to Germany for forced labor.

Many of the people of Le Chambon did not notice any sharp

change when the Germans occupied the Southern Zone, but Madame Eyraud did. Her house, situated at the angle made by two busy roads, seemed to invite the Gestapo and German military personnel, so that she always felt them "on my back." There were often German troops and vehicles passing by, but she had a clever way of managing things: she dressed her boys in inconspicuous clothing and let them blend into the passersby. She found this less dangerous than trying to confine the boys to the house, where a surprise raid could find them bunched together, nervous and suspicious-looking.

Her house was an especially dangerous one because many of the Maquis passed through it in the later years of the Occupation. Her husband, Léon Eyraud, was the center of an intelligence network that connected the Maquis, de Gaulle's Secret Army, and the BBC. He had a powerful radio receiver-transmitter, and occasionally a broadcast would come in from London: "Père Noël, Père Noël, gifts will be coming on Friday the twelfth." The word Noël is also *Léon* spelled backward, and the broadcasts were aimed at Léon Eyraud, so that he could look out for the next parachute drop of supplies for the various resistance troops in the region. One night about six months before the end of the war, a truck arrived loaded with arms for "Père Noël"—Madame Eyraud's boardinghouse was a busy place.

Madame Eyraud had been raised a Catholic, but during the war she had no religious affiliation, nor did her husband. In fact, religion played no part in their lives. As far as the hundreds of refugee boys who lived in her house were concerned, the only thing that counted was that the *gosses* (kids) should feel that they were at home in her house, completely at home, as if they were all her children.

This intimacy was to cause some pain at the end of the war. When the war was over and relatives of the boys came to find them in Le Chambon, the relatives discovered that the boys were deeply attached to the people and the countryside of the village. Many of the relatives who came were not parents—often the parents had been massacred—but grandparents, uncles, and aunts. The children had been with Madame Eyraud for an important part of their lives, and they felt closer to her and Le Chambon than they felt to these near-strangers who claimed them as relatives. Many of the relatives were destitute, and their poverty, combined with their anguish over what to do with their lives, now made this coldness, or at least this detachment, in the boys especially painful to bear for them.

The Marions had a boardinghouse not far from Madame Eyraud's. It sheltered adolescent girls, almost all of whom were Jewish, and most of whom were students in the Cévenol School. The words Madame Marion uses to describe life in her *pension* are very similar to the language Madame Eyraud uses: she wanted life in her house to be as much like normal French family life as it could be—with intimacy, relaxation, little chores to be done by all. When the war was over, her girls would bring back husbands and children and stay with them sometimes for weeks in order to enjoy a place where they had been so warmly welcomed in what they later discovered to have been such terrible times.

The Eyrauds had no interest in religion or the history of religion in France, but the Marions were Huguenots, and for them what happened in France and in their house was a reliving of the suffering of the Huguenots centuries before. When I asked them why they cooperated with Theis and Trocmé when the leaders brought them down this dangerous path of resistance to Vichy and later Nazi orders, one of the Marion daughters said, "What they were asking us to do was very much like what Protestants have done in France ever since the Reformation. Pastors were hidden here in Le Chambon from the sixteenth century through the period of the 'desert' in the eighteenth century. What we did was *le reste,* the traces of what was being done here for centuries."

Still another *pension* was that of the tiny, strong Alsatian, Madame Gabrielle Barraud. Her husband was Swiss, and a Communist, and did not take readily to all those nonpaying guests Madame Trocmé was sending to their *pension.* He was a deeply generous and gentle man, but there were limits to what he would do for capitalistic Christians. And so, as Madame Barraud puts it with that intense, closed look of hers and that gleeful twinkle in her eyes, "I would not let my left hand know what my Christian right hand was doing." She simply concealed the fact that certain of their guests—all boys—were nonpaying. Of all the heads of *pensions,* she was the closest personally to the Trocmés. Once she encountered Trocmé in the village square, and he looked down at her: "Oh, Madame Barraud, I would kiss you if I could, but you are so small, and my back hurts so that I can't bend! Maybe we can find a box to put you on!" Whenever she met either Magda or André Trocmé, after the first greeting she would ask, "And what do you need me for today?" Magda was always quick with an answer, with her unending store of tasks to be assigned, but André would sometimes say, "You're right! I talk to you now only when I need you!"

One of the reasons that she was in such close touch with the Trocmés was that her *pension* was very near the presbytery, and the boys she kept there stayed for only a little time in her house because the center of the village was the most dangerous part of Le Chambon. The frequent comings and goings of the boys kept her in almost daily communication with the presbytery, and this contributed to the danger in her situation. Once when I asked her to tell me her feelings about the danger of arrest and deportation, she said, "The Trocmés, you see, were dangerous people to follow. They wanted to help people in need by any means possible, and often those means were risky." When I asked her why she continued to help them, she said, "Oh, well, it was a matter of conscience. Whatever they asked for was just what my conscience would want. Why—I just could not have done anything else but help them and the refugees!" And when I pressed her on the question of conscience, she said that the most important influence on her thinking was the story of the Good Samaritan and the commandment to love one's neighbor as oneself.

Even though the boys stayed in her house only a short while, she, like all the heads of *pensions* I have talked with, wanted the children to feel as if they were her own. Once I asked her if she preferred some children to others. She told me that some were very well behaved, and some brought forth deep feelings of love in her, while others did not—but, she insisted, "I did all I could to keep the boys from knowing which ones I found more lovable."

Once she had in her house a very mischievous boy, who would not only steal but would even throw stones at her and others. When she asked him to excuse himself for doing those things, he refused. She told him he must leave her house and go to another one, and he chose to go to the House of the Rocks, one of the two houses directed by André's bespectacled cousin Daniel Trocmé. The boy was sixteen years old and Jewish. He went to the House of the Rocks just before the Gestapo raid that brought about the death of Trocmé and almost all the boys in the house. He was deported and died with those boys. Madame Barraud, whose daughter, Manou, was accidentally killed by one of the boys in her *pension,* lives a great deal in the past; she shrivels with remorse when she thinks of the story of that boy she could not continue to love.

There was another boy she remembered very clearly. He was Jewish and an epileptic. When he saw his father in prison in Paris, he had his first epileptic seizure, there before his imprisoned father. The father, who was a doctor, watched his son in the throes of the attack as he himself was being dragged out of a prison cell.

The boy loved to be loved, and Madame Barraud had all she could do to keep from lavishing more love upon him than she did upon the others. There was another fifteen-year-old in the *pension* who teased the epileptic boy a great deal. One day she took the stronger boy aside and told him that the epileptic lad was his brother, just as they were both her children. The teasing stopped.

After the war, the epileptic boy took up medicine and studied it for three years. One day when he was alone in his room at the university, he fell into a fit and died. He had been her darling, though she had done her best not to let the others know it.

She never went to sleep without knowing exactly where each boy was sleeping. One day she was musing aloud about a boy in her *pension* who had often kept her up late at night waiting for his return, and who had at last passed the *baccalauréat* examinations that opened the way for him to study medicine in the universities of France. He had been a nonpaying guest, and she had kept his spirits high during the exams. "Well," she said aloud to herself, "I've done one good thing in my life." One of her boys overheard her and said, "Among others."

There were more than a dozen such *pensions* in Le Chambon, but there was a greater number—and a number harder to ascertain—of private homes in which children and families stayed. Some of these homes kept people for short periods of time, but those on the outskirts of the village often kept refugees for years.

There is, for instance, the home of Ernest Chazot, which is on the road to the neighboring Protestant village of Le Mazet. His long-faced, energetic wife once stood on her front stoop and confronted a troop of Vichy police with the words "There are no refugees in here! And if there were, I wouldn't tell *you!*" But they had with them for more than three years a woman and her son from Vienna who were trying to find a life outside the control of Germany.

Some of the private homes were miles away from the village, on farms, a great number of which were owned by the Darbystes, the most fundamentalist of the Protestants in the region. Refugees would stay on the farms for long periods of time because of their safety, but also because of the special sympathy the Darbystes had for Jews. Believing that every word of the Bible was inspired by God, the Darbystes had a thorough knowledge of the history of the Jews as that history is told in the Old Testament.

Once, early in the Occupation, a German-Jewish refugee came to a Darbyste farm to buy some eggs on the unrationed "gray" market of the distant farms. She was invited into the kitchen. Quietly the woman who had invited her in asked, with the light of interest in her eyes, "You—you are Jewish?"

The woman, who had been tortured for her Jewishness, stepped back trembling, and she became even more frightened when the farm woman ran to the steps leading upstairs and called up, "Husband, children, come down, come down!"

But her fright disappeared when the woman added, while her family was coming down the steps, "Look, look, my family! We have in our house now a *representative of the Chosen People!*"

4·

Outside the ring of houses in our target pattern of the rescue activities of Le Chambon would have to stand the Cévenol School and its residences. Refugee children became part of its student body and refugee adults became part of its faculty during the war. The school was quite effective under the leadership of Édouard Theis and others at camouflaging their presence, so that during the whole Occupation there was not a single raid by either the Gestapo or the Vichy police on its classrooms or residences. One of the reasons for this was the location of the school in the wooded countryside north of Le Chambon, outside the mainstream of life in the village. But another reason was the quiet stability and orderliness of its faculty and students.

If the Cévenol School was quiet, the exact opposite was to be found in the groups in the last ring of our diagram: the Maquis and the Secret Army of de Gaulle. Those groups were dedicated to the liberation of France by violent means. In fact, so single-minded was that dedication that it could be argued that those organizations should not appear on a diagram of rescue activities in Le Chambon. The reason they are there—on the outermost circle of the diagram—is that they opposed the government and German occupants of France, as the nonviolent Chambonnais did, and they approved of the rescue activities but wanted to do *more:* they wanted to defeat the forces that made those rescue activities necessary, defeat them by force of arms. Moreover, in the last weeks of the war they protected Le Chambon by driving off marauding German squads.

Aside from the question of violence, there were irritating differences between them and those who were helping refugees. Many of the Maquis were headstrong, even wild. One day they "requisitioned" the stock of a tobacconist in Le Chambon and thereby aroused the anger of most of the village. They even angered the leaders of de Gaulle's Secret Army in the region, who were on the verge of shooting them down as looters. André Trocmé had to intervene and persuade the Maquis to leave Le Chambon.

Once a Catholic priest came to Trocmé and told him that his motorized bicycle had been "commandeered" by the Maquis. The priest knew of Trocmé's relationship with various students of the Cévenol School who had taken to the bushes despite Trocmé's beliefs but who were still close friends of the Trocmés. In dismay the priest said, "Pastor Trocmé, could you tell me why they have to use my motor-bicycle to save France?" Trocmé sought them out and returned the machine.

The housepainter Ernest Chazot, in whose home a Viennese mother and son stayed for years, had his little car stolen by the Maquis, and he learned that a teacher in nearby Yssingeaux had sold the car. Chazot could not reconcile this with patriotism.

Once some Maquisards who had been students at the Cévenol School and members of Trocmé's parish asked him to lend them plates and Communion cups so that they could have services in the field. They intended to form a "Christian Maquis," and they wanted to use the silver-plated cups and plates in the woods for Communion. Trocmé said: "I am sorry, but this is impossible. Get cups and plates like the ones you need from private parties. How can you reconcile Communion with the desire to kill Germans?" I have talked with one member of the Maquis who was angry at Trocmé for defining Christianity so narrowly.

Trocmé and his parish had a quieter relationship with the Secret Army of General de Gaulle. The leader of the local contingent was a Jew from Marseilles whom Trocmé admired profoundly. The leader tried to win the pastor over to the patient, moderate methods of de Gaulle's wing of the Resistance, but he understood and sympathized with Trocmé's reasons for refusing to join him, and came to visit Trocmé for friendly as well as tactical conversations. Though they differed on fundamental matters of conscience, they agreed on matters of practicality: they both believed that the very existence of Le Chambon depended upon maintaining peace.

"Père Noël" Eyraud, husband of the round-faced Madame Eyraud, who ran the most dangerous of the *pensions* of Le Chambon, was in some respects a bridge between the rescue activities of the village and the military or paramilitary aspects of the Resistance. Even though the BBC contacted him when military and other supplies were going to be dropped in the outskirts of the village, and even though the Maquis and members of the Secret Army were often to be seen in his home, the *pension,* he discouraged bloody moves against Vichy or the Germans, and prevented many a young hothead from trying to gun down German troops in the last years of the Occupation. He despised the cycle of revenge and would neither begin nor continue it, even though his function as the center of a network for intelligence-gathering had military aspects, such as the surveillance of German troop movements. He was conscious of the children his wife was sheltering and of the larger movement of which she was a part.

5.

There is a group within Le Chambon that does not appear on our diagram of the rescue activities conducted there: the Catholics in the village. A few Catholic families helped with rescue efforts, but most of the few dozen Catholic families in the town were not caught up in those activities. One of the reasons for this is that the priest of the small old Catholic church of Le Chambon was not interested in working closely with the Protestants around him. Another reason is that Protestants and Catholics in France at that time (and for centuries previous to that time) did not work or live closely together, except in rare instances.

All through the Occupation there had been local resistance against Vichy and the Nazis in various Catholic parishes, resistance that took the double form of refusal and help: refusal to do harm to Jews and other refugees, and help for them. One of the saints of the Resistance was Father Chaîllet, who, beginning in 1941, published one of the most important clandestine journals of the French Resistance, *Notebooks of Christian Testimony,* and who personally saved the lives of many Jewish refugee children in the region of Lyons. He had no political affiliations, and aside from saving children's lives, he wanted only to "awaken the French conscience," as he put it in the first issue of his *Notebooks.* Like Trocmé, he addressed his audience on a spiritual, not a political plane, urging them to say no in their thoughts and actions to what he called "Hitlerian thought." He was incarcerated in September 1942 in a psychiatric ward for refusing to follow Vichy's orders.

Perhaps the most eloquent of the Catholic leaders in France was Monsignor Saliège, archbishop of Toulouse. On Sunday, August 30, 1942, about a month after the horrific roundup of Jews in the Vélodrome d'Hiver in Paris, he announced his plans to read an important pastoral letter having to do with the Vélodrome d'Hiver roundup and the Jews having to wear the yellow star with the word *Jew* on it. The civil head of the region of Toulouse banned the letter, and various priests in the area spent Saturday night traveling to the home of their archbishop to ask whether they should now read the letter. Up to that time, high-ranking leaders of the Catholic Church in France had supported Pétain and his National Revolution against atheistic Communism. And so they needed help in dealing with this new policy on the part of the church hierarchy.

When they appeared at his door that Saturday night, he was asleep. Awakened and standing in his nightclothes, he called out, "They are to read it! They are to read it!"

In almost half the churches of the diocese of Toulouse, his pastoral letter was read. It is the single most eloquent document on Trocmé's and Chaîllet's "spiritual plane" of the French Resistance. It reads:

> Brethren:
>
> There is a Christian morality and a human ethic which impose duties and recognize rights. Both rights and duties are parts of human nature.
>
> They were sent by God. They can be violated. But no mortal sin can suppress them. The treatment of children, women, fathers, and mothers like a base herd of cattle, the separation of members of a family from one another and their deportation to unknown destinations, are sad spectacles which have been reserved for us to witness in our times. Why does the right of asylum of the church no longer exist? Why are we defeated? Lord, have mercy on us. Our Lady, pray for France. In our own diocese, in the camps of Noé and Récébédou, scenes of horror have taken place. Jews are men and women. Foreigners are men and women. It is just as criminal to use violence against these men and women, these fathers and mothers with families, as it is against anyone else. They too are members of the human race. They are our brothers like so many others. A Christian cannot forget that. France, our beloved country; France, known to all your children for a tradition of respect for human life; chivalrous, generous France, I trust in you and do not believe that you are responsible for these horrors.
>
> With my affectionate devotion,
> Jules-Gérard Saliège
> Archbishop of Toulouse

Magda Trocmé remembers that often phrases uttered or written by Saliège would become well known in Le Chambon. People would repeat his words to each other to raise their spirits. Once she told me:

> Oh, there were various monasteries and convents and Catholic homes where refugees were as safe as they were in Le Chambon, and we sent people there without hesitation. We never thought, "Those places are Protestant," or "Those places are Catholic." No. We thought only, "Those are people who will help."

Still, in the village of Le Chambon rescue activities and protests against the Vichy establishment were dominated by the Protestants. The tiny minority of Catholics was, on the whole, swept aside by the majority, which was being inspired and led by a Protestant minister whose temple and presbytery—not any neutral, secular ground—were the dynamo of resistance in Le Chambon.

Sometimes Magda and André Trocmé became conscious of the loneliness of the priest of the village, who had such a small parish to care for, and who, of course, had no family. Then they or other Protestants would bring him some special food that had come, perhaps, from a parachute drop (by way of the help of "Père Noël" Eyraud), and the *curé* would have a delicacy in those hard days. Or, in turn, he would catch a fine fish in the cool Lignon and bring it to the presbytery. But as a group, the Catholics of Le Chambon lived outside of the Resistance activities of Le Chambon.

6.

There is another group that does not appear on our target diagram of those activities: the refugees. One of the reasons for this is that the whole rescue network of Le Chambon was concerned with them. They cannot be put in any one ring in the diagram—they were everywhere in Le Chambon where help was being given. Another reason for their absence is that the diagram is supposed to convey the sources of initiative in the rescue activities. But though there were refugees in the armed Maquis around the village, the ones who came into the village itself, the "weak and oppressed," as Trocmé put it, were usually in no position to inspire or plan particular kinds of rescue efforts. Their distress initiated the rescue work of Le Chambon, but so overwhelming were the inimical forces around them that they had little room for maneuvering and less power for initiating special courses of action. Once at the Flowery Hill a woman, a Madame Bormann, fell on the floor and feigned madness during a *rafle,* so that the police making the raid let her be taken away to the village and to freedom. Later she told some Chambonnais that she had done this before to escape arrest. But such initiative was rare in Le Chambon, and became rarer and rarer as the occupants of France, the Germans, grew more and more desperately fearful of defeat.

There were various kinds of refugees in Le Chambon, and they shared only one important trait: they were all trying to escape from some form or other of Fascism. In the 1930s, the first refugees to come were Spanish Republicans, and they continued to arrive in small numbers throughout the years of the Occupation. One of them, called Pepito, figured importantly, as we shall see, in the story of the death of Daniel Trocmé. Aryan anti-Nazi Germans made up another group. Some of them, like the Hamkers, who lived with Ernest Chazot and his family throughout the war, left Germany or Austria to escape being mobilized in the German Army as well as to escape what they believed to be a vicious government. Still others left their homeland as an act of sheer protest against the government of Germany when there was no danger of their being mobilized or harmed by that government.

The refugees who came the latest to Le Chambon, but who grew rapidly in numbers as the Occupation went on, were the young men of France who were escaping forced labor in Germany. It was the Service du Travail Obligatoire and its increasingly stringent demands upon the young men of France that strengthened the Maquis more than any other act of the Vichy government. As the Germans grew to need more and more young men to work in their factories and fields while their own young men fought, Vichy and the Nazis made more and more demands upon the youth of France, and more and more of these young people either joined the Maquis or simply fled to some isolated place in France such as Le Chambon.

But the largest part of the refugees in Le Chambon were Jews. It was they, to use the formula that the rescue organization the Cimade used, who were "most endangered, and most dangerous to their hosts." At first, as the conversations between Trocmé and Chalmers indicate, Jewish children whose parents were in French internment camps or were being deported to German concentration camps were the main Jewish refugees in Le Chambon, especially in the funded houses; but early in the Occupation, Jews of all age groups came to Le Chambon, staying temporarily at the Flowery Hill or Madame Barraud's *pension* or the presbytery, or staying permanently at a private home like that of the Chazots or at the presbytery itself. In his notes, André Trocmé estimates that there were in the course of the Occupation about twenty-five hundred Jewish refugees of all age groups who came to Le Chambon. He could not document this number with records or any other evidence, and I have not been able to find any way of making a sound estimate of the number. But for the purposes of a study like this, the exact figure is of little importance. For us, the fact that individual human beings were saved is the most important fact. From the point of view of each of these refugees, Rabbi Hillel's statement in the Talmud is decisive: "If I am here, all is here; and if I am not here, who is here?"

As is the case with all groups of persons, the Jewish refugees differed from each other in various ways. All of them, especially those who stayed for periods longer than a few weeks, made themselves useful in Le Chambon, but at different kinds of tasks. The two refugees who spent the war in the presbytery, Madame Grünhut from Karlsruhe and Monsieur Kohn from Berlin, confined their activities to the busy presbytery itself. He was a cabinetmaker, and he spent much of his time repairing or constructing furniture. Madame Grünhut cooked, though her job was hard, since she had hardly any fats to cook with, and she found herself burning their already austere provisions. A Dr. Mautner from Vienna, who lived with Madame Barraud for a time, and who in his almost fiercely independent way wished to be of help, aided Drs. Le Forestier and Riou in their very burdensome medical practices. Still others worked with the Red Cross or on the faculty of the Cévenol School.

There were important differences aside from personal or occupational ones among them. Some were French Jews fleeing from the Nazi-ridden north in the early years of the Occupation. Many of these were shocked to be thought of as Jews and not simply as Frenchmen. People like Daniel Isaac, whose forebears were French from time out of mind, had not even thought of hiding their Jewishness before the war—their Jewishness had been of no great importance or even interest to them. The French they spoke was not identifiably "Jewish," nor was their behavior. Despite the infamous Dreyfus case at the turn of the century, for many of them anti-Semitism was something the Germans had brought into their lives in spite of their beloved country's efforts to prevent it. And as for Vichy, they believed that the marshal had been pressured and duped by the Nazis.

Most of them did not need new identity or ration cards, and there were often French friends who rallied round them. They were persecuted by government policies and by government-induced public opinion, and many were killed, but usually they were easier to hide and care for than the others. After all, they were still in their native land.

But the other Jewish refugees had deeper problems. The refugees fleeing from Eastern Europe had no allegiance to any country, or at least most of them had no profoundly felt allegiance to their host countries. They had lived in ghettos in the cities or in *shtetlach* (small towns or villages) in the countryside as exiles within the boundaries of their own countries. They thought of themselves mainly as Jews, not as Poles or Russians or Rumanians. Anti-Semitism was a part of their way of life, like earthquakes for people living on a great fault in the earth. Their countries did nothing to protect them from catastrophes like pogroms—on the contrary, the governments of their countries condoned or even initiated their torture and destruction. French Jews might pray for the well-being of the more enlightened leaders of France, but if there were a prayer from the Jews for the leader of an East European country, it might go: "Dear God, keep him well—well away from us." They were as accustomed to anti-Semitism as the French Jews were to being French. They had depths of anguish the French-assimilated Jews might never know, but they also had depths of fortitude and a trembling, passionate sensitivity born of centuries of coping with oppression.

The third kind of Jew in Le Chambon brought a complex fear into the community. He was the assimilated German or Austrian Jew. Perhaps his father—or he himself—had fought for his country proudly and at great cost. But whatever the particular connections with his country might be, they were often deep, and now he found himself in Le Chambon, with or without his family, not only wrenched from his country but hated and persecuted by that country as only the Nazis could hate. He was not even a human being they wanted to execute; he was one of the vermin to be exterminated. For many of these Jews, the tearing away from their homeland had been gradual, like the slow pulling off of a fingernail from a thumb.

One German refugee spent a year in the presbytery, helping the Trocmés to keep the children quiet and occupied, teaching them German, and helping hang up the washed bedlinen, which was changed every day of the week. She was the daughter of a respected teacher in Heidelberg, the university city that German poets have cherished for its physical beauty and its deep civilization. She had lived happily in an affluent part of the city until her friends started walking to school without her because they were afraid or ashamed to be seen with a Jewess. They stopped inviting her to their houses out of fear or shame or both until she had only one friend left: a girl whose parents owned a big hotel in the neighborhood. The hotel had a back door through which she could quietly enter in order to play with the daughter of the hotelkeeper. But soon the painfulness of their relationship grew too great and she stopped coming. The last Heidelberg friends to pull away from her family were the Catholics, who were also being persecuted. "They tried till the end," she said a third of a century after the event in her sweet, whimsical voice.

Her father, who taught in a distinguished secondary school, experienced the wrenching in other dimensions. He saw the Nazis removing the classics from his curriculum and replacing them with books, courses, and meetings whose sole purpose was the glorification of Germany. And he saw the Nazis putting into educational practice their basic philosophy of the profound unity between the Aryan people and the soil and the profound disunity between the sacred *Volk* and those floating disease germs, the Jews.

Suddenly he was arrested and sent to the concentration camp at Dachau, near Munich. But just as suddenly he was released from the deadly camp: he was needed to teach and administer his school. When he returned, friends and colleagues greeted him with a podium laden with flowers. There were older people in Germany, unlike the increasingly fanatical and barbaric young, who still had respect for classical learning and human decency. All their respect vanished when the power of Nazism in Germany reached its height in the late 1930s, and then his family found themselves refugees in France, rushing from one internment camp to the next, from one horror to the next.

His teenage daughter found herself in the presbytery toward the beginning of the Occupation. One winter night when the children were not feeling well, she spent the whole night going from child to child, rubbing each one down and making each a little more cheerful. The next morning, Trocmé found out that she had stayed awake all night doing this, and so he walked up to her and with that warm, wide smile of his, said, "You know, you are a real Christian."

Now, one of the ways of using the word *Christian* is as a term that separates one mode of worship from all others, and this meaning was much in use in her Europe. When she told me about this incident, I asked her whether she felt any resentment at having to be called a non-Jew in order to be praised highly. She knew exactly what I meant; many Jews who have been praised with the epithet *Christian* have felt a pang of bitterness at having to be removed from their religion in order to be praised, a pang that is made more painful by the obvious fact that the ascription is, in a plain sense, false: the praised Jew is still a Jew and not a Christian.

But when I asked the question, she responded with a clear smile and told me that for the Trocmés the word *Christian* had nothing to do with being a non-Jew. For them it referred to someone who loves his or her neighbor not with empty, self-indulgent emotion but helpfully, in action. In her year in the presbytery she grasped without any theological language what Trocmé later called the "basic truth" of his life: that there are no important divisions between human beings such as "Jew" and "non-Jew." The main distinction among people is between those who believe that those in need are as precious as they themselves are, and those who do not believe this.

One of her main tasks was keeping the children occupied without hindering the rapid, complicated activities of their parents. She hung out the sheets in the morning, tried to teach German to her beloved Daniel, who hated it, and piano to Nelly, who loved it (and especially loved that loud "Turkish March"), and supervised the home life of the children before little Jispa arrived.

Blossoming as she was into full womanhood, she was deeply interested in children. In the course of the year she spent in the presbytery—the first year of the Occupation—she saw how important the saving of refugee children was to the Trocmés. She felt that they wanted to save them because not only were the Trocmés "very, very concerned with helping and doing good," but they felt that "life was breathed into the children to make them *live.*" Looking back on this year, she remembers now not only how "very, very concerned with helping" the Trocmé parents were, but also that they were "always in a hurry, always in a hurry . . . dedicated. . . ."

7.

Once, in conversation with me, Magda Trocmé found herself comparing what happened in Le Chambon during the war with what the Cimade was then doing all over southern France. The Cimade, she said, was an organization with one clear goal and a definite organizational structure that reached down from international organizations into the teams for spiriting refugees across the natural and human obstacles between Le Chambon and neutral Switzerland. Madeleine Barot and other women had created and led it. But who had created Le Chambon? Magda Trocmé's husband had led it in his innovative way, but he had not created this little Huguenot community. It used its own deep resources to help the refugees. He had inspired it, but he had not made it what it was, the way Barot and the others had made the Cimade.

And saving refugees was *the* function of the Cimade; it was the only reason for its existence. But what was *the* function of Le Chambon, even during the war? Living, maybe; but to say this is to say something too vague to allow one to compare the two groups meaningfully. And Magda Trocmé went on, in her own breathless way: "You know? Saving refugees was a *hobby* for the people of Le Chambon! Oh, yes!" she called out in her slightly raucous, Italian-accented voice. "It was a hobby in Le Chambon."

Of course she did not mean a hobby in the sense of a pastime or a source of fun. No one knew better than she what sacrifices and what dangers the Chambonnais had taken upon themselves when they accepted refugees into their houses while they were all "under the German cup." She meant a hobby in the sense of something done by an untrained amateur, a lover of the thing done, not a professional, skillful practitioner of an art or science.

She meant that the people of Le Chambon saved refugees on the side, so to speak, as an avocation, not as part of their vocation, as a way of saving lives, not as a way of making a living.

When a refugee comes to your door and you say, as Magda Trocmé did, "Naturally, come in, and come in," you are not inventing an institution, nor are you participating in one that has already been invented. And when the refugees start arriving in larger and larger numbers (like a drop of oil, spreading, with no stages in their increase) and the one o'clock afternoon train that brings them is more and more full, you are still only a person, not a rescuer, only a person accepting people from the outside into the very center of your home. Your vocation is perhaps that of a carpenter, a painter, a housewife, or a farmer; it is certainly not that of a rescuer.

Trocmé's decision to make a place of refuge in the funded houses was only a part of what happened in Le Chambon. Before that decision, refugees were sheltered in the village, and after it, many different kinds of ways were devised for helping the refugees: the *pensions,* the farmhouses, private homes in the village proper, and the residences of the Cévenol School. Le Chambon became a village of refuge not by fiat, not by virtue of the decision Trocmé or any other person made, but by virtue of the fact that, after Magda Trocmé's first encounters with refugees, no Chambonnais ever turned away a refugee, and no Chambonnais ever denounced or betrayed a refugee. In addition to the burdens of their harsh life during the Occupation, they found themselves taking on the burdens of the constantly increasing danger and the constantly increasing hunger. There was no one decision that made Le Chambon a place of refuge, the safest in Europe, but an attitude expressed in French as being *toujours prêt, toujours prêt à rendre service* (always ready, always ready to help).

For this reason, the target diagram we have been using to summarize the rescue activities in Le Chambon is somewhat misleading. Inspiration and guidance did emanate from the presbytery and the temple, but the village became a place of refuge more by accretion than by planning. The rings of our diagram were far from being as closely and neatly related to each other in Le Chambon as they are in the diagram itself. In the end, each private home, each *pension,* even each funded house simply coped on its own.

We cannot understand the rescue activities of Le Chambon if we do not realize that the village was far from being an organization with neatly meshing parts. We can understand them only if we see how discreet those rescue activities were. They were *discrete* in the sense of being separate from each other, distinct, and they were *discreet* in the sense that they were silent, even cautious with each other. Miraculously, there were no chatterboxes or gossips or boasters or frightened complainers—at least none that I have been able to discover—among those who housed the refugees. Magda Trocmé, with her immense powers of speedy speech and extensive knowledge of what was happening in the houses of Le Chambon (the most extensive, perhaps, of anybody in the village except her husband), was, during the Occupation, utterly silent about those activities. A Chambonnais once told me, "You cannot make a place of refuge with a bunch of talkative people." The Chambonnais practiced their avocation in secrecy.

For example, Madame Marion, who had a boardinghouse for girls, might see a light in the house next door in a room that had been vacant for months. Ah, she would think, I shall have a little joke with Madame Russier [or some other common Chambonnais name]; I shall ask her who is there. The next time she encountered her, she would ask Madame Russier whether a relative had come to visit her. Madame Russier would look at her half-humorously and say, "Well, Madame Marion, I thought I should clean up the room, you know, air it out a bit in the cool evening air. . . ." And both would know, though neither would say so, that another refugee or another family of refugees had come to Le Chambon.

Now the French, especially in the villages, are a very private people, as the windows tightly shuttered at night suggest, and as

Laurence Wylie has shown so memorably in his *Village in the Vaucluse.* But discretion during the Occupation was not simply a result of custom; it was a felt necessity. As few people as possible had to know the whereabouts of a refugee or a refugee family because the greater the number of people who knew, the greater the probability that an idle word or gesture would trigger an arrest and deportation, and the greater the probability that severe questioning by the authorities would cause somebody to give damning information about another.

If you possess a dangerous bit of information, there is usually a limit to the amount of resistance, the amount of secrecy you can maintain, especially under the influence of a very powerful adversary. For most human beings, there is a point beyond which silence is impossible when they are subjected to torture or even trickery. But if people are ignorant of certain dangerous facts, facts dangerous when they are revealed to an adversary possessing great power, then as far as those facts are concerned, there are no limits to the resistance those people can exert. There is nothing for them to reveal, and so there is no danger of their revealing anything. Those who do not know a secret are the best guardians of that secret. An awareness of these considerations led various leaders of the French Resistance who knew many dangerous facts to commit suicide before they reached that point when they could no longer resist divulging the knowledge they had. If they had been ignorant of those facts, their silence concerning them, their silent resistance, would have been endless. And so the Chambonnais wished to know or divulge to each other as few facts as possible concerning the refugees.

Though they were discreet, as silent and as separate as possible regarding the refugees, the amateurs of Le Chambon had a sense of fellowship with each other in the face of the suffering they were helping to alleviate. Many of them attended the temple sermons of Theis and Trocmé and participated in the weekly discussions with the *responsables* of the parish. Because of such sessions, and because of a feeling that they were, in the words of Madame

Eyraud, "doing something of consequence" *(quelque chose pleine de conséquences),* there was a sense of fellowship not entirely dissimilar to the feeling of fellowship at the camp at Saint-Paul d'Eyjeaux when Theis, Darcissac, and Trocmé were imprisoned there. It was this feeling of fellowship in doing, again to quote Madame Eyraud, "something good" *(quelque chose de bien),* that made the Chambonnais come so swiftly to the presbytery on the night that the leaders were arrested, and that made them sing to Trocmé as he walked to the police cars.

On the surface in ordinary times during the Occupation, when there was no visible crisis, the lives of the people of Le Chambon showed few signs of such solidarity. Magda Trocmé once said: "If it had been an organization, it could not have worked. How can you have a big organization deciding on people who were streaming through houses? When the refugees were there, on your doorstep, in danger, there were decisions that had to be made then and there. Red tape would have kept us from saving many of them. Everybody was free to decide swiftly on his own." Again the important element of time—only independent individuals could act swiftly and surely in those circumstances. Le Chambon in the first four years of the 1940s was a world in which delay in letting somebody come into your house amounted to a decision *not* to help in that all-important *now.*

What unified and divided the Chambonnais then was what we might grasp by way of the metaphor of "atmosphere." People shared a dangerous and helpful atmosphere the way people share physical air; it was all around them, and yet each drew it into his or her own life, the way we share and divide the air we breathe.

This intermingling of separateness and communion is nowhere more clearly exhibited than in the important activity of making identity and ration cards for foreign refugees. The cards could be made only from fresh blanks; old cards with former impressions of a governmental seal could not be converted. Somehow these blanks turned up in the houses of the people of Le Chambon when new refugees arrived. Everybody I have asked—and I have asked heads of *pensions,* heads of private homes, and even leaders like Roger Darcissac and Magda Trocmé—gives me a different account of where they came from. One suggested the Cimade; another was fairly certain that a secretary to a mayor in some town hall—either the town hall of Le Chambon itself or the town hall of Le Mazet, or perhaps the city hall of Le Puy, the principal city of the region—supplied them; still another suggested that a Jew living in the nearby town of Tence secured the cards and somehow had them sent to the right Chambonnais for distribution. And still another suggested that they came from nearby Yssingeaux. Magda Trocmé, among others, did not claim to know who supplied them, and she, like the others, never asked. For her, the blanks turned up in her kitchen near the entrance from the little hallway that opens to both the kitchen and the dining room, or they appeared deeper in the house, sometimes in her husband's office. Since the doors to the presbytery were not locked, it would have been easy for any one of many people to place the cards in the house without being noticed.

She, like the others who never asked, felt that knowing where they came from was both unnecessary—they almost invariably arrived in time—and dangerous for all concerned. Despite this ignorance, the cards continued to come throughout the Occupation. Somebody knew and somebody cared about each new refugee and about each new group of refugees that came to Le Chambon, though for most of the Chambonnais there was only silence on the subject. This efficacious silence is a symbol and an example of the intermingling of privacy and communion that pervaded Le Chambon during the rescue activities.

PART FOUR

Consequences— 1943–1944

8

Daniel Trocmé, a Conscience Without Gaps

1.

If you stand firmly opposed to overwhelming destructive power, you expose yourself to destruction. And the risk you run becomes especially great when you are not part of an organization staffed by trained soldiers issuing and obeying clear, circumspect commands. The Chambonnais were nonprofessionals. They had not been trained to do what they did. Each of them was an amateur at saving lives, both in the sense of the Latin word *amator* (lover) and in the modern sense of the word *amateur* (one who is not trained to make his living in a given occupation). Under the leadership of Trocmé, they expressed love for their fellow human beings, and they expressed it not through discipline but through an alert common sense.

Scattered through their several houses in the village itself and the farms surrounding it for miles in all directions, they followed no battle plan, but instead waited for someone to knock at their doors—a refugee, an agent of Vichy or the Germans, or a friend. Despite the Sunday morning sermons, the meetings of the thirteen *responsables,* and the talks in the dining room and office of the presbytery, the "kitchen battle" of Le Chambon was more reactive than initiatory. It was a series of day-to-day responses to initiatives that came from the outside world. There were procedures for safeguarding refugees (first reception, then the printing of necessary cards during the temporary lodging of the refugee, and finally placement in an outlying house or with a Cimade team for the trip across the mountains to Switzerland), but throughout these procedures, each Chambonnais had to act entirely at his or her own discretion. A soldier in a theater of war or even in a bush battle has the security of following the commands of a superior and of using the training he himself has received—his duty has been made clear to him by others, and usually well in advance of the action. The Chambonnais had little such security; in the end, they had only a dangerous refugee in their house, and their own scruples.

But scruples can be not only unclear and idiosyncratic; they can be imprudent as well. The word *scruples* comes from the Latin word for "pebble." Scruples, like sharp stones in a shoe, can hinder a retreat from danger. Moral scruples tend to make one think of others, instead of thinking of one's own safety. They are unlike the skills of a hardened soldier, whose combat hardening consists in learning how to stay alive while helping his unit to survive and prevail. In the battles of World War II (as in all violent battles since the beginnings of combat), risk hardened people, made it safer and safer to work with them as they acquired more experience—if, of course, the risk was not so great as to unman them. Usually, to go into battle with a veteran by your side was far safer than to go with a "green" recruit whom risk had not honed to a sharp cutting edge.

But risk gave the scrupulous Chambonnais no such skills. Over the years of the Occupation they did not so much learn how to take care of themselves as that saving refugees was an important activity. Increasing danger taught them that the refugees were really in mortal danger and desperately needed their help. If the Gestapo threatened and even killed an unarmed person for protecting a refugee, what would they not do to their prey once they caught him? The increasing risks of the "kitchen struggle" did not harden the Chambonnais; it did not make them more and more safe by making them more and more canny at staying alive. It convinced them that they were needed by the refugees. The hardened soldier learns how to take care of himself and his buddies; the Chambonnais learned that they *must* take care of those others, or they would surely die. Risk made them set their jaws in the service of life; it did not make them safer soldiers. It made each one more of an amateur.

Magda Trocmé has told me that all the dangers that grew steadily greater as the Germans grew more and more defensive and desperate in the last months of the Occupation helped the Chambonnais to concentrate on saving the refugees. She knew and cared nothing about how those risks might have helped harden the Chambonnais by teaching them how to protect themselves from arrest.

All of this meant that as the Germans grew more frustrated and more dangerous, the Chambonnais grew more vulnerable to brute force, since they were more interested in defending others than they were in defending themselves. And so there were deaths. One of the amateurs who were killed was Daniel Trocmé.

2.

After Trocmé's "easy victory" with his presbyterial council in the winter of 1941 over the question of making Le Chambon a place of refuge by establishing a residence in the village funded by the Quakers, he asked his second cousin Daniel Trocmé to take over that residence. Here is how Trocmé describes his slender, intense cousin in his notes: he was "an intellectual given to rather vague ideas, and often rather absentminded, but totally free of selfishness, and possessed of a conscience without gaps." His dreaminess might make handling young adolescent boys difficult for him, but it was mainly his conscience, his devotion to others, his capacity to think not in terms of his own security or comfort but in terms of the welfare of the children, that drew André Trocmé to him.

His father was a first cousin of Trocmé, and was head of the famous École des Roches, one of the finest secondary schools in France. There were eight children in the family, but they adopted one more because they had "good educational facilities" for raising children.

In the winter of 1940–1941, Daniel was in the southwest of France, teaching in his father's school, which had been moved there before the Nazis struck the north. He was restless, in fact resentful at the comparative luxury of his life in his father's school, and he responded to his cousin's invitation swiftly and eagerly.

He had a heart ailment that made it dangerous for him to do strenuous physical work, especially high in the mountains, but once he was in Le Chambon as monitor of the Crickets, he worked, to use Magda Trocmé's phrase, "like a madman." He was in his mid-twenties, his nerves tensely strung, his eyes behind those steel-rimmed spectacles always demanding. At the Crickets, and soon afterward also at the House of the Rocks, he spent his evenings resoling the shoes of the children with old automobile tires and making a vast tureen of soup for them to eat the next day. At lunchtime he appeared in the schoolyard dragging his wagon with the metal tureen upon it, and he saw to it that each of his children got plenty of the hot soup. Then, with that weak heart of his, he dragged the wagon with the tureen on it north up the steep hill on which the village square sits, and then, farther north, to the Crickets, which was even farther away from the square than the Cévenol School.

Once, during a *rafle* in the square in the third year of the Occupation, he saw a Vichy policeman smiling as people were being put into a bus—he had worked hard for this moment. Suddenly the amateur Daniel Trocmé secured the attention of the people in the square and said loudly, pointing at the Vichy policeman, "I accuse this man of being responsible for this arrest." As the bus left the square, he led the Chambonnais in singing their old song of farewell to friends, which had become a song of the Resistance. Sung to the melody of "Auld Lang Syne," it went:

> It's only *au revoir,* my brothers;
> It's only *au revoir.*
> We shall be together again, my brothers;
> We shall be together again.

A hardened bush-warrior told me this story, and he also told me that it is quite possible that this action led to Daniel Trocmé's arrest and death. And the soldier went on, his eyes wet, "They had no training, these people, no discipline. You know, they were *imprudent!* They did not seem to care that they were dealing with a great and murderous force!"

In the summer of 1943, when the jonquils were gone but the violently yellow-gold *genêts* were swaying in their long-fingered, multifarious way, the Gestapo made their only successful raid upon a funded house of Le Chambon. They struck Daniel Trocmé's House of the Rocks. It was one of those days in Le Chambon that had made the village a tourist center. The air was so clear that the vast, distant extinct volcano to the south, Le Mézenc, had a visible granular texture. On such a summer day in Le Chambon, the whistling of a farmboy kilometers away struck the ear with its original breathy freshness.

Magda Trocmé was without most of her family in the presbytery. They and their two permanent refugee guests, Madame

Grünhut and Monsieur Kohn, were away on vacation in a house lent them by friends. She and her son Jean-Pierre had come back early to Le Chambon in order to help some of the older students of the Cévenol School prepare for their important *baccalauréat* examinations, which they were to take in the nearby city of Le Puy.

Into her kitchen rushed a young woman named Suzanne, a seventeen-year-old who was at the school on a scholarship, and whose family life was so full of pain that the only adult she could love and trust was Daniel Trocmé. "Come, Madame Trocmé!" she called out as she entered the kitchen. "The Gestapo have had a roundup at the Crickets, and they have taken Daniel and the children to the House of the Rocks. You know, he could have escaped through the back door of the Crickets—the woods are close by there—but he felt responsible for the children in his two houses, and so he let them take him."

The Crickets was mainly for children between eight and twelve years old, but it housed a few younger and a few older ones at that time. Daniel had a special concern for this refuge full of very young people, and aside from bringing them soup every lunchtime and repairing their shoes, he had been patching their clothes day and night. Funded though it was, this house, like the others, was very poor, like the commune of Le Chambon in which it stood. In those days, almost everybody except soldiers walked around in patches. And as if he were not doing enough for the children, he occasionally rushed off to the prefect of the region to complain that there was not enough oil or fat for cooking.

The moment Suzanne finished telling about Daniel's arrest, the pastor's wife walked straight out of the kitchen, took her bicycle from under the roof of the wooden porch that opened onto the street, and drove to the House of the Rocks, where the Gestapo were holding them all. As she came up to the portal of the driveway leading to the three-story granite house, she saw that there were three buses parked in front of the main entrance. Later she found out that two other Chambonnais had already tried to get in and had failed: a local doctor and a teacher in the Cévenol School. Now all she knew was that she was going in, and she was not going to be stopped. And sure enough, to her surprise, she walked right between the German guards into the house. Later she realized that she had been admitted because in her haste to go to Daniel's and the children's aid, she had forgotten to remove the apron she had been wearing in the kitchen, and the guards had taken her for a maid. This is a plausible explanation; with her brusque way of walking through obstacles without hesitation, her big, brown braids atop her head, and her strong features and sturdy body, she must have been taken—by a stranger to Le Chambon—for a maid who belonged in the house.

When she stepped into the dining room, she saw all the young people of the house sitting with their backs against the walls. In the middle of the room there was a table that had been placed there by the German secret police. Sitting at the table were four or five armed civilian Germans and one Chambonnais, the man who was in charge of purchasing food for the house. Magda smiled inwardly: the Germans had chosen the wrong person to represent the house. Daniel, the monitor of the house, was sitting surrounded by students in a corner against a wall, apparently quite relaxed.

Without thinking, carried along by her smooth entry into the house and room, she walked straight up to Daniel, and as she did so, one of the Gestapo men screamed at her in German. ("You know how the Germans like to scream," she said later.) Understanding German and realizing that she had almost made a fatal mistake by identifying Daniel as the head of the house, she abruptly turned and went into the kitchen that adjoined the big dining room. There, sitting at the kitchen table, was the German refugee cook who had been hired by the Cévenol School to feed their own students and everybody else in the House of the Rocks. The cook must have been surprised to see the pastor's wife suddenly appear in her apron and casually join her in the kitchen. But without breaking her stride, Magda sat down at the table with the cook and said nothing. Both sat there in silence.

It was about ten o'clock in the morning. About an hour after she entered the house she heard calls in German from the Gestapo in the dining room. They wanted eggs and bread and something to drink. And so she and the cook made two eggs and bread for each Gestapo agent and each German guard. Glancing at each other with joy at the windfall, the two women made the same food for themselves, and they ate it swiftly before the Germans could stop them. Usually they had one egg—or at most two—a month.

After they had eaten—nobody else had been fed—one Gestapo agent crossed the kitchen and went into a little room with a large notebook under his arm. One by one the young people in the house who had been sitting around the dining room crossed the kitchen on their way to the little room, and the agent interrogated them and checked their names against a list in his notebook. The door was closed.

Some of the students came back through the kitchen with bruised faces. Despite the quietness of the interviews, this was no casual interrogation. As each one crossed the kitchen near Magda Trocmé, either going to or coming from the inspection room, he found a moment to say a few words to the person who was now in charge in Le Chambon. One said, "Here is money. Send it to my mother—you know the address." Another said, "I have a watch in my room. Send it to my family." Another said, "Write to my girl friend. Here is her address."

When Daniel came out of the interrogation room, Magda believed that he had been held there a long time because not only was the name Trocmé on a special list of suspects, but he spoke German fluently, too fluently for a Frenchman, and they suspected him of being a German Jew. Their suspicions had grown stronger when under severe questioning his conscience without gaps had made him defend the Jews. If he had been a soldier, he would have been trained and committed to reply briefly and minimally to interrogation by the enemy. But all he had was that conscience. Even prudence could not make a gap in it.

After the interrogation, the Gestapo decided that they would give the prisoners something to eat. Magda and the German cook used tiny containers to bring the bits of bread and water to them, and Magda tried to soothe as many of them as possible while she was passing out the bread and water. During this lengthy procedure, she had a chance to talk at some length with Daniel. He told her that a few weeks before, one of the boys, a refugee from Franco's Spain, had saved the life of a German soldier who was drowning in the Lignon.

The young Spaniard was called Pepito, but nobody in Le Chambon seems to remember his whole name. But Miss Maber, the teacher of English at the Cévenol School, remembers that he had been chosen as the best all-round athlete in the universities of Spain.

Daniel suggested that she go immediately to a high-ranking officer of the German troops staying at the Hôtel du Lignon and plead not only for the release of Pepito but also for the release of the rest of the prisoners on the grounds that Le Chambon had been kind to a German soldier.

As soon as she could, Magda left, got back on her bicycle, and went straight to the Hôtel du Lignon, the best hotel in the village, where German Commandant Headquarters were located. On the way—the trip to the hotel was not a short one—her meticulous mind sorted out the problems now before her. She had to get into the officer's rooms without having to deal with inferiors, who would slow her down and possibly stop her; she knew about the German policy of laying full and exclusive responsibility upon a leader in all decision-making. And she had to find a German witness, or at least a credible one, who could verify Daniel Trocmé's claim about Pepito.

Having definite goals, and being a healthy, rather handsome woman in her early forties, she stopped her bicycle as daintily as possible at the curb outside the main entrance of the hotel. She may have minced, but only as well as Magda Trocmé can mince, which is not very well, as she approached the guard standing before the door. She had long ago learned that with men, to play the lady was to play a winning card. But when she asked the German soldier to allow her to see the senior officer, he answered firmly, "*Verboten.*" She had also learned that if she raised her head high and demanded—always in a ladylike manner, of course—to be treated by the man facing her in a gentlemanly way, she usually got what she wanted. This she now did, and the intimidated young soldier let her in, with instructions for reaching the officer's rooms.

When she arrived at the right floor, she saw two or three officers standing near the rooms of the senior officer. Deciding to change her plan of action to fit the circumstances, she walked straight up to them and asked how long they had been stationed in Le Chambon. When they expressed surprise at her abrupt question, she told them that she had a message to deliver, but she could transmit it only if she knew when its recipients had arrived in Le Chambon. As she had planned, the officers grew curious, and when one of them told her that he had been there well over three weeks, she asked if he was aware of the rescue of a German soldier from the Lignon River. The officer said that he was, and that he knew that a young Spanish refugee had been the one who had saved the soldier's life.

Her heart rising, she said to him, "The Gestapo is at the House of the Rocks. Please, please come with me and repeat what you have just told me about the soldier's rescue. Many lives depend upon it."

One of the officers said, "We have nothing to do with the Gestapo."

The pastor's wife replied, "I am a lady; you are an officer. Being a witness is your simple duty as an officer and a gentleman. It is simply a matter of honor."

"How did you get here?" asked one of the officers.

"By bicycle," she replied.

The officer said, "Well, then, go down and get it, and we'll follow you there in a little while."

But time was too precious to be lost now, and besides, she was not so confident of the officers' sense of honor as she had said she was. "No," she said. "Come with me; we shall walk together."

And so it happened that on a fine, gold-flowering summer's day, Magda Trocmé, one of the key figures in the Resistance of southern France, was seen walking down the streets of Le Chambon with a young German officer on her right, and a middle-aged German officer on her left, like any collaborator.

As they left the village, she saw approaching them two young ladies from the local Young Women's Christian Association (one of the Protestant groups in Le Chambon). The looks on their faces when they saw her approaching with those uniformed Germans was, she said later, an interesting sight to behold. There is an appropriate French word for describing their condition: *médusées* (petrified, as if they had looked into the face of the Gorgon Medusa). They blinked as if to try to wipe out or test the image in their heads. And when Magda asked them to hand over their bicycles to the two officers, their faces froze again into utter immobility. "But," said Magda Trocmé later, "they had a kind of respect for me, you see, and they loved me, yes, and so they gave us the bicycles."

Mounted upon women's bicycles, the three rode east on the road to Saint-Agrève through the waving golden *genêts,* which are very thick between the village and the House of the Rocks. But Magda was not one to waste valuable time with chitchat or silence or gazing at scenery. Calling out to the officers (it was hard to speak casually while calling out from a moving bicycle to someone else on a moving bicycle), she mentioned the deplorable news of Gestapo agents killing people in nearby Clermont-Ferrand. She reminded them that these killings of university people (the faculty and administration of the University of Strasbourg had been moved to Clermont-Ferrand before the Germans took over southern France) had made a bad impression on the French people and on the rest of the world. And then she mentioned other killings recently perpetrated by the Gestapo.

Pumping away, one of the officers responded, "Well, this is war. When one of our men is killed, an example must be made."

Now Magda could take the next step in her little plan. "But if all that is war, how do you explain why you people in the army have to go and fight the Russians under great danger on the Eastern Front, while the Gestapo stay safely at home to fight *their* war? How do you explain this?"

The younger officer, being quicker than the older one, assumed an expression that read like this to Magda: Well, this is no stupid woman. She's making sense. But she is also trying to divide and conquer us by making invidious comparisons between the army and the Gestapo.

The older one said, "All the Gestapo people have been on the Russian Front."

She was not hoping to arouse any powerful feelings against the secret police in the bosoms of the officers; all she was hoping for was that one of them might think: Well, I'll tell that Gestapo agent a thing or two. After all, a German *soldier* was saved.

As they pulled up to the wide gate of the House of the Rocks, she decided that there was nothing more she could do than to introduce the officers to the Gestapo agents, mention Pepito and the rescue of the German soldier, and leave after a brief talk with Daniel. She managed to do all these things, and Pepito was released.

When she returned to the presbytery, she found her eldest son, Jean-Pierre, waiting for her. He had turned fourteen in April, and was as intense as any of the Trocmés. She told him the story, and that she had to go back to Daniel's house to find out whether they had released the boys and Daniel. He said that he wanted to accompany her on his own bicycle, and that he would not let her go alone. She knew, and so did he, that it was more dangerous for him to be near the house than it was for her, since he was the same age as the students who had been arrested. She tried to dissuade him, but he insisted. She was afraid and touched when, after doing a few chores, she found him waiting in the porch with her bicycle and his.

As they approached the gate to the driveway of the house, they saw a bit of Nazi Germany: on a staircase outside the house, a blond, blue-eyed Jewish boy who was one of the residents of the house was crouched upon the bottom steps; above him a Gestapo agent was standing with one hand grasping the black iron railing of the staircase, and with the other he was whipping the boy across the head and shoulders with a set of phylacteries. A Jewish boy usually receives his phylacteries after his Bar Mitzvah, when at the age of thirteen he is admitted to the Jewish community as a full-fledged member. In the hard boxes of the phylacteries are Hebrew texts, usually on vellum. Donning these phylacteries by using their leather straps, he prays with them, and they remind him to obey the Law. While the Gestapo man was beating the boy with the instruments of prayer, he was screaming down at him, *"Schweinejude! Schweinejude!"* ("Pig-Jew! Pig-Jew!") It was an act of pure hatred, with no practical police motive in it.

At first the pastor's wife and son thought he was beating the boy with a belt; when they learned what those leather strips and those boxes were, they stood there stiff with horror. Magda had to hold fourteen-year-old Jean-Pierre back because he was straining to interfere. Such an act would have doomed them all.

Once through the gate, she approached Daniel as he stood at the head of the line waiting to enter the buses. She was going to ask him what the possibilities of release were, but he started talking before she could speak. He said, "Write immediately to my parents. Tell them I'll send them news as soon as I can. Tell them I'm okay, and not to worry. And remind them how much I like traveling."

These words, about his pleasure in traveling, uttered when he knew he was in great danger, were the last any of the Trocmés would ever hear from his lips.

When they had all left in the three buses, Magda turned to her son, who, like his father, had a streak of violence in him. The boy's face was flushed with anger and bewilderment, and his eyes were brimful of tears ready to spill down his cheeks. "Mother," he said, "this is too terrible. Such things should not exist. As soon as I am big enough, I am going to have *revenge.*"

"But, Jean-Pierre, you know your father's ideas. You know that there is nonviolence, that there is another way of handling things."

"Oh, Mother," he said, and never finished the reply.

3.

A year later, one of the young men who had been taken away came back to Le Chambon and told the Trocmés what had happened to Daniel. He had been questioned frequently all the way to the killing camp at Maidanek in eastern Poland. The one topic that the interrogations emphasized was his attitude toward the Jews. Again and again he expressed his compassion for them, until one of his questioners openly said to him, "You must be one of them—otherwise you could not defend them so."

Then began a long search for the young man, a search that ended after his parents were both dead, when word came from the Russians who had captured Maidanek that the name of Daniel Trocmé was on the list of those gassed and incinerated there. In the late spring of 1945, it was finally confirmed that Daniel Trocmé had been killed on April 4, 1944, at two o'clock in the morning at the death camp of Maidanek in Poland.

Even after his death, the Gestapo continued to investigate his Jewishness—they wanted to be certain that he had committed the ultimate crime, the crime of being Jewish. In May 1944, the city hall of the city in which Daniel had been born received a letter from the Gestapo asking them to verify the Germans' belief that Daniel was in fact a Jew.

In 1976, Israel awarded him the Medal of Righteousness posthumously. As they had done for André Trocmé, they planted a tree in his memory. Now there are two Trocmé trees in the Holocaust Memorial of Yad Vashem in Jerusalem.

9

Flight from the Gestapo

1.

It was in the middle of the summer of 1943, when the lucid air of Le Chambon made gold flowers and green stalks so vivid that the colors seemed palpable. Pastor Trocmé was sitting in his office in the presbytery when a boy in patches—like all the other boys in Le Chambon—rushed in crying, "Monsieur Trocmé! They've just shot Praly!" Trocmé pulled his body out of the chair, with that twinge of pain that was always in his back, crossed his office and the dining room in a few long steps, and pounded his way up the crooked Rue de la Grande Fontaine to the Hôtel des Acacias, where Praly was dying.

Commissioner Praly was not a Chambonnais: he had been sent to the village by Vichy; he was young, handsome, and a Protestant, and he was there to gather information about the enemies of France who were using Le Chambon as a shelter. Some of those enemies were armed—usually with hunting rifles. They were the ones beginning to be called the Maquis. As yet there were not many of them. But the other enemies of France—the Jews—were there in considerable number.

Worshiping in the Protestant temple and drinking in the cafés, smiling, clear-eyed Praly was in Le Chambon to obtain by patient listening, money, and threats the names and locations of the bush-warriors and Jews in the area. Every weekday evening he stepped out of the Hôtel des Acacias, which stood opposite the railroad station, and, without trying to conceal what he was doing —the young man had style—he mailed to Vichy a big envelope stuffed with information about this troublesome commune.

One day early in the summer, he and the pastor happened to meet in the street. Dropping his usual mask of lightheartedness, Praly warned Trocmé about one of his homes for refugee children, the House of the Rocks, which was headed by the pastor's second cousin, Daniel Trocmé. He told the pastor with severity that he should do something about this "dangerous shelter for Jews and antipatriots." This was the only hint the Chambonnais got of the raid that was soon to come. In response, Trocmé asked Praly why he had taken on this terrible job of informing against the poor and the persecuted. Deftly putting on his usual mask of gaiety, Praly looked up at the minister and answered, with a sly wink that assumed that Trocmé would understand, "Well, Monsieur Trocmé, you know each of us must earn his living as best he can." Half-humorous as it was, that remark was all there was to be said as far as he was concerned.

Now he had been in Le Chambon almost a year, and the larger groups of guerrilla warriors living around the village (Trocmé had convinced them to live and to conduct their activities outside his nonviolent village) had decided that Praly was not the light-headed clown he appeared to be, but was a solid danger to the Resistance. And so, on that midsummer day, they had sent four of their number to eliminate the danger.

Two of them stood guard outside the hotel while the other two went inside and asked Praly, who was dining, to step into the hallway outside the dining room. When he did so, they drew their revolvers and shot him. Then they coolly joined their companions, who had been holding their bicycles for them, and rode away from Le Chambon. Trocmé never learned their names.

The pastor arrived while Praly was being carried on a stretcher to an ambulance. Plainly, he was dying. The minister bent over him and said gently, "Monsieur Praly, you have been doing a sad task. And you have suffered a terrible punishment for it. God will pardon you for what you have been doing if you will ask for His forgiveness now."

The idea of repentance was important to Trocmé; one of his main reasons for advocating and practicing nonviolence was that he wanted people to give their enemies the chance to repent, instead of killing them before they could do so.

The gray-white-skinned young man closed his eyes and kept his silence. He had said his last word on the subject in the streets of Le Chambon, with that wink. He bled to death in the ambulance on the way to the nearby city of Le Puy.

In the months that followed, the pressure on the village from both Vichy and the Nazis became heavier and heavier. The killing of Praly had helped—in all likelihood—to bring about the raid on the Crickets and the House of the Rocks. Praly had warned the pastor about "that dangerous shelter for Jews and antipatriots," the House of the Rocks. Now the whole commune was regarded as such a shelter.

And so it was no great surprise to Trocmé when, in July of that eventful summer of 1943, a young member of the Maquis walked into Trocmé's office and told him that there was a price on the pastor's head.

He claimed to be a double agent for the Gestapo and the Maquis, and told Trocmé that he had attended a meeting of the Gestapo in the nearby city of Valence, and had learned there that the German secret police had decided to have Trocmé assassinated. He was not to be arrested and deported because such an act against a man as conspicuous as Trocmé would arouse too much resentment against the Germans at a time when they needed all their manpower and equipment to keep the war from being lost. He was to be killed by a hit-and-run murderer in the employ of the Gestapo, but not identifiable as somebody connected with them. The young man urged Trocmé to go into hiding.

Trocmé was reluctant to go for various reasons. As an advocate of nonviolence, he had been accused often—mainly by himself—of the fear of violence, of cowardice. He felt that this new danger might well be a test of his courage to face violence when it was directed against himself. But more important, he knew that in Le Chambon, and indeed in the region at large, he was an example of successful nonviolent resistance, and he was afraid that if he ran away from the Nazis instead of standing and resisting them, the whole spirit of nonviolent resistance in the area might crack, and all the acts of refusal and help that the villagers and peasants had been performing would end. He was afraid that Le Chambon might be a Célisse—needing him and suicidal without him. And so he hesitated to follow the young man's advice.

When a high official of the Reformed Church of France, who was also a trusted friend, came to talk with him, his doubtfulness increased. The official said that Le Chambon had had enough martyrs, including Daniel Trocmé and the children in his charge, who might all be dead by then. He went on to say that the village might well turn away from nonviolence if Trocmé were to be killed—the Chambonnais were already deeply disturbed over the arrests and the rumors of possible additional roundups. The official went on, "You know what happens in these assassinations. You're picked up in a car without warning and your body is found in the woods, or the French criminals the Gestapo hires to do their dirty work burst into your house during a meal when everybody is at the table and spray bullets everywhere. Will you allow not only yourself but your wife and children and the refugees to be killed?"

Trocmé answered that he would not, but added, "Look, if I

leave the village, there will not be a strong voice here for nonviolence, and the village may let itself turn violent."

The official answered, "You will be away only for a few weeks. The BBC says that there will be an invasion this summer! You'll be free then." He and the BBC were wrong—the landing on the beaches of Normandy was almost a year away—but of course Trocmé did not know this.

Trocmé was still torn. He wondered how he would explain to God his running away from the forces of evil—God, who had permitted Jesus to be crucified by those forces.

Finally, on a bright Sunday when the slender, now flowerless *genêts* were swaying by the roadsides, the whole Trocmé family rode out of Le Chambon on bicycles in the first stage of Trocmé's plan to escape. Typically, Trocmé gives no decisive reason for his decision to flee—his notes simply list the reasons for his doubts. In the book he would write during the ten months of his flight (the book would not be published), he would explain that in the face of stubborn contradictions, such as that between his duty to stay in Le Chambon and his need to leave, a person cannot hope to dissolve the contradiction by neat, clear logic. In such situations, one must simply *bet* upon a certain course of action—one must, in an act of faith, throw oneself into action in a certain direction. For instance, when confronted by the conflicts between the methods and findings of modern science on the one hand and the stories in the Bible on the other, one must make a leap in one direction or the other—there is no neat theory to reconcile the two or to tell us which is "right." One must do almost an act of violence to one's own rational tendencies—one must leap, bet, *act* before it is too late.

For Trocmé, there had been no such contradictions involved in his refusal to obey Vichy and the Nazis, and there had been no such contradictions involved in his labors on behalf of the refugees. These actions came straight from a life-and-death ethic that God, through Moses and Jesus, had commanded men to follow. But the decision to leave Le Chambon, like the decision to choose Christianity with its Resurrection instead of modern science, involved launching oneself into action in the face of hopeless contradictions.

Now his leap, his bet, gave him confidence that, the decision to leave having been made, all would turn out well. As always, his power to bet, to throw his whole mind and body into what he was doing, free of debilitating doubt, gave him energy and firmness.

On the road east to Saint-Agrève, he left his family. A coolly brave hardware dealer picked him up in an automobile, leaving him at nightfall in the Protestant presbytery of Lamastre, where he was given a room. Magda had had a ration card made for him and an identity card under the name "Béguet," with a picture of him without a mustache and wearing his glasses; he had shaved his mustache, acquired the glasses, and gotten Roger Darcissac to take the picture before he left Le Chambon.

The pastor of Lamastre was on vacation when Trocmé arrived. When he returned, he angrily sent the dangerous Trocmé away. But it was time for him to go anyway: some Protestant working girls had seen him one day getting some fresh air at the window of the presbytery. And so one evening the brave hardware dealer picked him up again and took him to a place situated near a high hill from which members of the household or Trocmé himself could see any police cars that might be approaching.

It turned out that the move was a good one, one of the many lucky moves in Trocmé's life. Soon after he left the presbytery in Lamastre the Gestapo came to the place looking for him. Not being able to find him, they lost his trail—they found no informers in Lamastre.

Trocmé spent many weeks in the house by the hill. He was now in the department of Ardèche, adjoining his own department of Haute-Loire. Especially in those hard times, the basis of the diet of the Ardèchois was chestnuts, and he spent his evenings peeling chestnuts. Thanks to eating so many of them during his enforced rest inside the hill house, he put on a great deal of weight, most of which he was never able to lose. For the rest of his life he would be a rather stout man, carrying extra weight upon his painful back.

After a while, to keep his trail unclear, he moved into a country house whose only other inhabitant was the daughter of the owner. She was in her mid-thirties. When she started to ask him to do little, intimate things for her, like arranging a cushion under her head, he sent a letter to his wife by way of a third person asking her to come to help him out of a compromising situation —his long aversion to extramarital sex relationships had made the call necessary, even though traveling was dangerous in the region, and even though it was especially dangerous for the wife of the pastor to leave Le Chambon and go to him. A few days later, along came Magda on her bicycle. She had ridden dozens of miles on roundabout roads, including a few miles in a dried-up creek bed, to get to him without being followed. There she stood before him now: thinner than ever because of the meager food of Le Chambon and the immense amount of work she was doing there. But she was there, ready, as always, to help him out of a confusing situation that was too delicate to discuss with any other person.

In a few days, he was at another home in the Drôme, this time a boardinghouse blessed with a very fine cook. He grew even heavier with weight he would never lose. So sedentary was his life there that he never left the house even for Sunday services. Unable to resist the temptation to eat much good food after all that deprivation, he was becoming stouter, putting more weight on that ruptured disc in his backbone.

After a few weeks in the *pension* he learned that his son Jacques was longing for him so much that the boy was not able to do his schoolwork. Trocmé sent for the boy. With the help of a Jewish intellectual who lived in the neighborhood, and who taught Jacot mathematics, he undertook the education of his son. What followed was one of the happiest times in Trocmé's life. Winter had come, and the two threw snowballs at each other, played at cowboys and Indians around and inside the house, and tried to raise a litter of mice they had found there. The mice died, but the two Trocmés were thriving joyously together, living in the same big room, Jacot with a small desk, and his father with a big one. The son remembers after more than a third of a century how swiftly their friendship became sweet and deep, with no parish, no brothers or sister, no mother, no anything to come between them.

Then came the event that almost killed them both and that reveals the soul of the "soul of Le Chambon" as vividly as any other event in his life could do. Trocmé had come to Lyons for a vital meeting, and had to leave the city at eight-thirty one Sunday morning on the train for Valence. They arrived at the great railroad station rather too close to eight-thirty, and they still had to pick up some luggage from the baggage room before boarding the train. In order to do this in time, Trocmé left Jacques with their hand luggage in front of the station so that he could pick up the bags unhindered.

He was running toward the baggage room when he heard cries behind him; but he did not think the cries were directed at him, and so he kept on running. Suddenly he found himself standing face-to-face with a violently angry German soldier who was pointing his pistol straight at Trocmé's chest. Behind Trocmé stood another menacing German soldier. Frightened, and not knowing the cause of all this, Trocmé stood still. The soldier in front of him pushed his gun hard against Trocmé's chest, so that he had to step back; in stepping back, Trocmé found himself falling backward into a sitting position. He looked around and saw that he was sitting in the rear doorway of a prison van. The two soldiers picked up his feet, shoved him inside the van, and locked the door. He was arrested by the German police.

Sitting in the barred vehicle, he mustered his thoughts. He must have been arrested for running in the station, not because he was on the Gestapo death list; nonetheless, he was being arrested—he had been caught. Soon the German police would interrogate him and demand to see his papers. His identity card gave his name as Béguet, and they would ask him if this was indeed true. Then he would have to lie in order to hide his identity. But he was not able to lie; lying, especially to save his own skin, was "sliding toward those compromises that God had not called upon me to make," he wrote in his autobiographical notes on this incident. Saving the lives of others—and even saving his own life—with false identity cards was one thing, but standing before another human being and speaking lies to him only for the sake of self-preservation was something different. *Telling the policeman a lie face-to-face* would mean crossing a line that stands between the false identity card that saves a human life and the betrayal of one's fellowman and of one's God. Trocmé had allowed the false identity card to be made for him only to give sympathetic French police an excuse for not turning him over to the German police. And in Le Chambon the false identity cards *for others* were a weapon against the betrayal of those who were being persecuted. Trocmé had never lied to Vichy or to the Germans about there being refugees in Le Chambon. He had told them frankly that there were, and he had just as frankly refused to tell them who and where they were. He had defended the defenseless, and he had not betrayed the people in his charge. Now he would be bearing false witness, lying simply to save himself.

He decided that when the German police questioned him, he would say, "I am not Monsieur Béguet. I am Pastor André Trocmé." Having made this decision, he became calm; his conscience was quiet.

But a thought shattered his calm: Jacot was waiting for his father in front of the station, and he did not know his way back to their friends in Lyons or to Le Chambon. He was a sensitive, very emotional boy, deeply attached to his father. How his horror would grow as the hours passed and his father did not return!

And so Trocmé's agile mind took another tack. Through the barred window in the rear door of the van, he called out in German, "Listen for a moment!" The guard did not answer. Glad of his knowledge of German, Trocmé persisted, and at last the guard appeared at the barred window.

"Look," Trocmé said in his fluent German, "I have my child, a boy of twelve years, blond, tall, who is waiting for me with our baggage in front of the station. Please bring him here; I've got to tell him where he must go, since I am arrested."

The guard was skeptical and told Trocmé to repeat his story to his immediate superior, who in turn called their captain to the window of the van. Trocmé repeated his story to the captain, who then said, "You're making up all of this. Why were you running?"

"I was late for the train to Valence. I was running to pick up my baggage. Look—here is my baggage check, and here are our tickets."

"You mean you were not running away from the roundup that's going on here?"

"What roundup?"

"Didn't you notice that the whole station is surrounded by police?"

"No. I was in a hurry to catch the train."

"You crossed a ring of police without noticing it, then? That's not a likely story." Then the captain thought awhile and added, "Let's see how true your story is." He told the guard to take Trocmé to the front of the station, and if his son was there waiting for him, to bring them both back. Then he turned to Trocmé and warned him that if he was trying to trick the police, or if he was going to try to escape, he was a dead man.

After he let him out of the van, the guard shoved the muzzle of his gun between Trocmé's shoulder blades and yelled, *"Los!"* ("Take off!") With the gun in his back, Trocmé marched off like an automaton toward the front of the station. As he walked, he thought, If only Jacques is still there and has not panicked . . . Then he saw the boy waving happily at him from a distance.

Jacques Trocmé, after all those years, remembers that moment clearly. He had been frightened at the disappearance of his father, but at his absence, not because of any fears about his well-being. When he saw his father coming, even though he had a soldier behind him, and even if that soldier was going to shoot him, he felt pure joy: at least his father would be with him now; at least he could see him.

But the idea that his father was in danger was a fleeting one. According to Trocmé's notes, the boy might have thought that the soldier was helping them with their luggage! Once, in Valence, Magda Trocmé had commandeered a German soldier to help them with their luggage. Magda, as always, was ready and eager to take as well as give help. When her husband scolded her for her boldness, she had replied, "Good heavens! They have to be good for *something!*" And so the boy might have been thinking that his father had learned a lesson from his mother and was bringing a German soldier along to help them. He did not see the gun rammed into his father's back. All he knew was happiness at his presence.

But soon the gun came into sight, and Trocmé said, "I am arrested, Jacot. Come with me so that I can explain where you should go from here."

"Papa! Papa! What's going to become of us?" cried the suddenly desperate twelve-year-old.

The three returned to the captain, and the officer studied the beautiful blond child leaning on his father, broken with despair and horror. The German officer was deeply, visibly moved. "You are telling the truth, I see," he said to Trocmé. Then he told the guard to take them both to the control point in the station, but to keep a close watch on them.

The guard put them in a long line that led up to an officer sitting at a table. He was carefully examining the papers of each person in the line, and he was consulting a small directory that obviously contained the photographs of people wanted by the police. He was not only checking documents; he was also comparing the faces of the voyagers with the photographs.

Trocmé knew that the ordeal was far from over. His papers were false and had not been made with great care; besides, his name was Béguet on those papers, and though he had said that the boy with him was his son, the name on Jacot's papers was

Trocmé. Moreover, even with his son there, he would not try to lie his way out of arrest.

It was necessary to avoid passing through the control point, but the guard who had put them in line was standing close by, watching. How could he do it?

Because of the care the officer was using on each person, the line was moving slowly. A bit bored, the guard started talking with his buddies stationed nearby, and as the line moved, he neglected to walk alongside the Trocmés. Trocmé decided that if he could put the stone pillar that now stood before him between himself and the guard, he could find out whether the soldier was watching them very closely. If he was not, then they could make a move to avoid passing through the control point.

"Jacot," he murmured to his child, "do exactly what I do, slowly and very calmly. Keep your baggage in your hands while you do it."

"Yes, Papa," the boy said.

The guard was not looking at them. The two left the line with a few steps and were soon behind the pillar and out of his sight. There was no audible reaction from the soldier.

Trocmé had noticed that the control was for departing voyagers only, not for people arriving in the station in the adjoining area. He had also noticed that the exit from the station was not far from the pillar behind which they were now standing. Walking sedately, he told his son, "Let us leave this area slowly." When they got to an exit from the section for departing passengers, they entered into the stream of arriving passengers, descended the tiny flight of stairs near the main door to the station, and stepped out of the station and onto a tram that went by it just as they heard loud yells and whistles coming from the railroad station: the Germans had discovered they were gone just a moment too late.

A quarter of an hour later, they were singing a hymn in a Protestant church, for it was Sunday, and they were free—free, Trocmé emphasizes in his notes, without having lied to anyone.

Good luck—happy events that have their origin outside of our control—had not deserted the Trocmés. In one of the very few times in the history of the German occupation of France, somebody had been arrested, had been put in a prison van, and had been allowed to leave without a single policeman or soldier thinking of making that fateful, often-heard demand, *"Papier! Papier!"* If anyone had asked for his papers, Trocmé, and quite possibly Jacques as well, would have been deported and destroyed, because Trocmé would have told the truth.

From the Protestant church they returned to the home of their Lyons friend, Madame Paillot. The next morning they came back to the station with her. She was walking ahead of them, and they were walking separately from each other. If she saw that the roundup was still going on in the station, she would drop her handkerchief and they would go their separate ways back to her home. But she dropped no handkerchief, and the two entered the station and took the train for Valence.

As they were leaving the Valence station through an underground passage, they saw two German policemen, each wearing a metal crescent hanging from his neck. They were stopping departing passengers with the familiar *"Papier! Papier!"* As soon as he saw the two policemen ahead, Trocmé turned to his son and said rather loudly in German, "Why, we have taken the wrong passageway! We'd better go the other way." And they walked back up the stairs, out of the passageway, and out of a side door of the station.

Many years later, Trocmé asked himself: "How were we saved?" Part of the answer for him was that the tears of his child had touched the feelings of the captain. But the main and for him the surest part of the answer was God: "God did not wish me to die yet," he wrote in his notes. Apparently it never crossed his mind to take seriously the fact that his own cool observation, thinking, and action had done much to save them, just as it never crossed his mind to take seriously the possibility of lying in a face-to-face confrontation.

After the war, a professor of history came to visit the Trocmés in their little cottage near Geneva. He was there to interview them about the French Resistance during World War II. In the course of their conversation he asked Trocmé, "Exactly what happened when you were arrested at the beginning of 1944 in the railroad station at Lyons?"

"How could you possibly know about that?" Trocmé asked in surprise. "Nobody knew about it!"

"Far from it," the historian replied. "I have access to the files of the Gestapo in Lyons. Right after your escape you were recognized as Pastor Trocmé by a photo they had—but of course it was too late. You were gone. How did you do it?"

After Trocmé told him the story, he said, "Well, you certainly didn't lack coolness. Incidentally, the officer who let you get away was held responsible for your escape. According to the archives, he was sent to the Russian Front, where they sent police who were guilty of negligence."

Trocmé's response to this information was: "I hope that our good and unknowing liberator did not fall on the Russian Front. He wouldn't have deserved such a fate, because he was able to be moved by the countenance of a weeping child."

The end of Trocmé's flight came shortly after June 6, 1944, when the Allies landed on the beaches of Normandy. He and his son arrived in Le Chambon toward evening, and their reception was glorious. Magda, Nelly, Jean-Pierre, Daniel, the parish, and the village rejoiced at their return, and their joy was all the greater because it seemed then that the end of the war was at hand. But they were wrong; there was death and suffering ahead.

2.

While Trocmé had been away, another leader of Le Chambon, Roger Darcissac, had remained in the village. His name was not on the Gestapo death list, and besides, he had an obligation as an educator, as the director of the Boys' School of Le Chambon, to stay in place and keep life as normal as possible in the village. Not that he dropped out of the Resistance; far from it. As the last months of the war and the Occupation were approaching, he was making more and more false identity cards, and doing more and more to shelter the Maquis in his school, and to help in the distribution of the supplies being dropped by parachute from Free French and British planes.

But of the three leaders it was Édouard Theis who was leading the most dangerous, and perhaps the most useful, life during those ten months preceding the landings in Normandy. Because his name was on the Gestapo death list with Trocmé's, he had left Le Chambon a few days after the pastor. But instead of merely hiding, he took a more active stance: he became a *passeur-pasteur* (passer-pastor) with the redoubtable Cimade. He became one of the comparatively few male members of a Cimade team, whose main task was taking refugees to the frontier of neutral Switzerland.

At the beginning of the perilous trips to the Swiss frontier, leaders of the teams forbade any talking on the part of the refugees, except in utter privacy, because their German or Polish accents could destroy them and the whole team. Papers and instructions given, they would start their trip, moving from *gîte d'étape* to *gîte d'étape* (from underground railroad station to underground railroad station). They might spend one night at the Catholic abbey at Tamie, and another at the Protestant presbytery at Annecy, and still another at the Catholic presbytery at Douvaine, so intimately were the Catholics and Protestants of France cooperating with each other in the saving of Jewish lives (for almost all of those in great danger now and seeking asylum in Switzerland were Jews).

There was usually barbed wire on the French side of the frontier, at least in populated areas, and one priest had a large concrete drainage pipe put underneath the wires, apparently connecting the base of a hill in France to the base of a hill in Switzerland, but actually permitting the Cimade teams to crawl under the wire without being caught in it or seized by the French or German police.

As in all complex operations involving great danger, there were tragic events. A team might get to the Swiss frontier only to find that one or more of the refugees in the team were not on a Swiss list, or had their names misspelled on such a list. Then families might be divided, and one part of the team would have to go back to France in order to correct the Swiss list. Teams were caught and their members either deported or killed outright. Theis himself was under arrest for a while, but fortunately in Switzerland, and Protestant authorities working in Switzerland with the Cimade secured his release.

He returned to Le Chambon long after Trocmé did, after the Germans had all gone from Le Chambon, after the liberation of the area. He had worked for more than a year for the refugees with the Cimade.

I have asked him why, when he was trying to avoid arrest and death, he chose to put himself in very great danger with the Cimade. His response was typical—a long silence, and then a wry little inwardly directed smile that brought his big, pointed nose even farther down toward his chin, and finally some words: "Ah." Silence. "It was not reasonable. But you know, I had to do it anyway." There is no man I have ever met who is so reluctant to use the language of moral judgment, words like "right" and "wrong" or "good" and "evil." In fact he speaks reluctantly in any language.

I have walked with him down an icy road in Massachusetts as he escorted me to my hotel room from his daughter's house. And, being a Jew, I have imagined vividly how other Jews must have felt walking beside Édouard Theis through danger toward safety. They must have felt his massiveness, of course, but above all, they must have felt the utter simplicity of heart and speech of the "rock of Le Chambon." They must have felt that he would, without a word, risk everything to bring them to safety.

10

The Death of an Eccentric

1.

It was in 1939 that the brown-haired, slender, vivaciously handsome Roger Le Forestier arrived in Le Chambon. He had been working with Albert Schweitzer, the great thinker who had given meaning to his reverence for life not only in theology but also in his work in music and medicine. But Le Forestier (for reasons I have not been able to discover) had not been happy in Africa with Schweitzer, and he had come to Le Chambon in hopes of working closely with another man who was known to have a practical reverence for life, André Trocmé. One day in that year before the fall of France, he appeared in Trocmé's dark office, and he seemed to illuminate it by his presence. Full of enthusiasm, though saddened and physically debilitated by his work in Africa, he asked Trocmé if he could stay in the presbytery and serve as a doctor in the village. The pastor was delighted to have him in the village, but felt that if he lived in the presbytery, it would look as if Trocmé preferred him over the other doctor in the village, Dr. Riou. Besides, Trocmé went on to explain to Le Forestier, Magda was already too busy with her teaching and the parish and the presbytery itself, and a doctor in the house would make her work even more tiring, what with frequent phone calls at all hours and even more comings and goings in the house than she was already experiencing. And so they shook hands and Le Forestier left the presbytery to find another place to live and conduct his medical practice.

A few days later, the laughing, youthful man in his early twenties reappeared in the presbytery with his luggage and announced that he was staying *there,* after all. Magda remembers that he was a "lonely, young, sad, beautiful man, handicapped by his failure with Schweitzer." And so, irresistibly, he moved into the presbytery.

Some Chambonnais were shocked at his unbridled spontaneity, but almost everybody who met him came to adore him. In the presbytery the children played games with him, or sat at the dining-room table with him, imagining colorful worlds. Once when Magda came back from the hospital after an operation, she found them all around the table dressed in white, as if they were all doctors, and ready to eat their food not with knives and forks but with Le Forestier's surgical instruments. After the fall of France in 1940, almost every day he would put his head out a window overlooking the Lignon and yell to the countryside, "It is now high noon in London!" And soon after he moved into the presbytery (while the Trocmés were away), he painted the wooden walls of the dining room yellow and orange.

Early in the Occupation he fell in love with Danielle, a woman as handsome as he was. And his love for her was as enthusiastic as the rest of his life. Later Magda said, "My goodness, he couldn't stop being in love with that girl!" They married in her native city of Cannes, and they took Magda with them for the wedding and the early days of their honeymoon. They all stayed for a while in Monte Carlo, and the joyous Le Forestier charmed a band of musicians into following them on their picnics and through the streets, playing their instruments and singing. In the end, the musicians had to leave—they had to make a living, after all—and so they kissed Magda, Danielle, and Roger good-bye.

During the rest of their honeymoon, without Magda, they went to visit Schweitzer in Lambaréné, and returned with a monkey, which they left in the presbytery when they found a home of their own in Le Chambon. The monkey was attached to Magda, and was wildly jealous of the children; it fought them off when they tried to kiss her before and after school. But soon the monkey died—it could not take the cold of Le Chambon.

Those who spoke of Le Forestier a third of a century after the war spoke of him as if he were still alive; so vividly did he live that his death and his absence seemed less real than his life. Magda spoke of him without a trace of her customary brusqueness, but with wide eyes, and the Marions, who had an important boarding-house for girls during the Occupation, spoke of him with undisguisedly gleeful worship. Miss Maber, the English teacher at the Cévenol School, who was one of the people at the school most involved in helping refugee children, described him as "completely mad, a genius."

His genius, she went on, was in the main the genius of a distinguished surgeon. One day a young man bicycling down to the center of the village from one of the funded houses took his hands off the handlebars in youthful bravado, slid across the road, collided with a low wall, and flew over the wall onto a stone courtyard. His intestines were spread around him. Instead of simply sewing him up and telling the bereaved parents how sad it was that the boy had to die of infection, Le Forestier took him to a hospital in Saint-Agrève, washed his intestines there inch by inch, coiled them up again inside the boy, and sewed him up. Miss Maber said, "As far as I know, the boy is still alive."

But his love of life was not confined to his personal affairs and his profession. Immediately after France fell in June 1940, he joined the Resistance. He was totally committed to nonviolence, but as a doctor he was useful to the Maquis, who were often moving in harm's way. Like others in France, he knew that in the early years of the war there was no hope of expelling the German invaders from France or of overturning the Vichy government by means of local Resistance activities; the Maquis were too weak, too loosely coordinated, and too ill-supplied to make any real impact upon the leaders of France. But Le Forestier felt that resistance to what you believe to be evil was right and necessary no matter how ineffective that resistance might be. One evening he walked back and forth outside Ernest Chazot's house trying to convince Chazot that conscience demands even hopeless resistance, but the housepainter Chazot insisted that only after the Allies made a landing in France or Italy, only after the Germans were beginning to weaken, should there be active resistance—otherwise the resisters would be endangering their own lives and the lives of those around them in an empty, idealistic enterprise. Roger Le Forestier was a principled enthusiast; Ernest Chazot was a practical man.

But the Maquis were sometimes afraid of Le Forestier's enthusiasm. Whenever a committed professional soldier from de Gaulle's Secret Army or from Britain arrived in Le Chambon, they tried to keep Le Forestier out of his way. Once an English officer parachuted down outside the village, and upon meeting this lighthearted Frenchman, turned to the obvious leader of the village, André Trocmé, and asked him openly why such a person was involved with the serious business of saving France from Vichy and the Germans. Trocmé assured the officer that Le Forestier could be trusted despite his talkativeness and his sometimes wild sense of humor. Besides, Trocmé went on, he was the only doctor the Maquis had, and he was a good one. Actually Trocmé cherished Le Forestier for the doctor's fervent belief in nonviolence and for his desire to keep the Maquis from destroying human life. Like "Père Noël," Léon Eyraud, he was a force for good within the Maquis. After listening to Trocmé's defense of Le Forestier, the English officer looked the doctor over again grimly and suggested that Trocmé keep this figure of fun in his own office, away from the serious business of driving out the Germans.

In the summer of 1944, after the landings on the beaches of Normandy, the air was full of stories of massacres that the now defensive Germans were perpetrating. Villages that had Jews in them and that had members of the Maquis nearby were being attacked by parachutists and gliders, and massacred. The Resistance in Le Chambon felt itself surrounded. To the west, toward Le Puy, was the dreaded Tartar Legion, SS troops trained to kill not only soldiers but civilians; to the east, in the direction of Saint-Agrève, German troops were coming up the Rhone River valley supported by the French version of the SS, the Milice. Many indications—including the concentrations of Maquis and Secret Army troops around Le Chambon, and the presence of Trocmé's and Theis's names on the Gestapo death lists—pointed toward Le Chambon as marked for massacre.

And there was another source of danger: French patriots themselves. The transfer of power from Vichy to some other group was as frightening a problem now as the transfer of power in 1940 from a defeated Third Republic to Vichy had been. De Gaulle refused to take authority from Pétain's hands, and he was deeply suspicious of the Maquis. The Maquis, in turn, were settling their own obscure scores all over France with executions that could occur without trial. Pétain, with his policy of solidarity with his friends and enmity against his opponents, had deeply divided France during the Occupation; now as France was moving toward the Liberation, those divisions multiplied and often became bloody. With members of the Resistance quarreling with one another, and with all of them bitterly attacking those who collaborated with the Germans, France in the last few months before the Liberation was in a state resembling civil war—except that this kind of civil war involved hatreds and suspicions *within* each of the two major camps. The chaos was even more complex than it had been, for example, in the American Civil War. In *this* terrible time in France there was no clear north and south, as there had been in nineteenth-century America.

One day the news came that the Germans were approaching from the east, from the direction of Saint-Agrève, and it was decided that the village must be evacuated. The mayor of Le Chambon and all the other leaders of the village swiftly organized first an alert and then the evacuation itself. By noon the village was deserted. Magda and the children, with many others, were in the nearby farmhouses, and the Chambonnais were awaiting the arrival of the German troops.

That evening, Trocmé entered the village square and saw a group of Maquis arguing violently with Le Forestier. They wanted his ambulance so that they could transport troops to the eastern part of the plateau where the Germans were going to arrive. Le Forestier was refusing, on the grounds that the ambulance bore a large red cross, which, according to the Geneva Convention, indicated that the vehicle was not transporting either arms or soldiers. "Besides," Le Forestier was arguing, "I am a doctor for the Maquis. What if there are wounded men around here and they need me, and I have no ambulance?"

Trocmé defended Le Forestier and emphasized the illegality of using Red Cross ambulances for nonmedical, military purposes. One of the Maquis said, "To hell with the Red Cross!" and facing Le Forestier, he yelled, "Give us that key!"

Suddenly an English officer wearing the beret of a paratrooper pushed his way through the group, and with his heavy English accent asked what was happening. One of the Maquis yelled out, "This man is a traitor to the Maquis! Kill him!" The English officer, who had been living amidst crosscurrents of violent anger since he had set foot upon the soil of France, drew his revolver and pointed it at the chest of Le Forestier. He was ready to shoot. Only Trocmé with his English and his hard-earned prestige could stop him; but the officer demanded that Le Forestier be put under arrest for refusing to go into combat, despite Trocmé's insistence that the doctor was there to help the wounded who would come in from the areas of battle surrounding Le Chambon. In the end the English officer let Trocmé take Le Forestier back to the empty presbytery, with the promise from Le Forestier that he would not run away, and with the pastor standing surety for him.

Having left Le Forestier in the presbytery, Trocmé walked around the village. He learned that there was in fact a battle on the eastern edge of the plateau at Saint-Agrève, and that there was a need there for Le Forestier and his ambulance. He rushed back to the presbytery. Le Forestier was not there! He had violated his word.

To his utter amazement, Trocmé found the doctor at home, sitting relaxedly on the balcony with his family, sipping tea. When Trocmé told him that he should go to the front with the ambulance, Le Forestier refused. "I'm not going," he said, "after the way I've been treated! Me, a prisoner of the Maquis? Me, one of their first friends, and this damned Englishman was going to shoot me as a spy if you hadn't stopped him? Never!"

Nothing Trocmé could say made a difference. His final word was, "No, Monsieur Trocmé. I am staying here. I don't give a damn about their wounded or their battle. Besides, I don't even think there was a battle at Saint-Agrève." Disgusted, Trocmé left him.

It turned out that Le Forestier was right: there had been no battle at Saint-Agrève. The people in the region had panicked after the Germans and the Milice had attacked the town of Le Chaylard, but the enemy troops had turned away to the Rhone valley. That evening the Maquis came back, ashamed because they had no battles to report, and the people of Le Chambon left the woods and had supper in their houses.

But Le Forestier, even though he had been right that there had been no battle at the edge of the plateau, and even though he was a lighthearted, unpredictably whimsical man, felt a burden of proof upon his shoulders: he had to prove that he was no coward for having stayed in Le Chambon. Soon he would furnish that proof, but at the cost of his life.

2.

A few days later, during those last months before the liberation of southern France, Le Forestier's wife, Danielle, came to the presbytery to ask for help: "Roger has got it into his head to bring about the release of two Maquisards who are in prison at Le Puy." When Trocmé asked him about this, he told the pastor that if the Maquisards stayed in prison much longer, the Gestapo might well deport them, and so some of the Maquis had decided to storm the prison at Le Puy and release the two prisoners before this could happen. He went on to say that storming the prison was an idiotic idea, and he felt that it was his duty to plead with the authorities in Le Puy for their release before their fellow Maquis could strike.

When Trocmé begged him to stay off the dangerous roads—Germans were shooting at vehicles that were not their own—Le Forestier replied, "I've been accused of cowardice about the Saint-Agrève business. And when there is a meeting of the Council of Resistance of Le Chambon, they do not invite me anymore, even though I was its founder! They want to take away my list of the members of the council! But I'm going to keep it."

Trocmé was dismayed at these words. Not only was he going to try to prove himself by driving to Le Puy on those dangerous roads and pleading with the prefect for the release of Maquis prisoners; he was keeping a list of the leaders of the local Resistance in his house, even though it was an unbreakable principle of the Maquis (and of the Secret Army) never to keep in one's home lists of even its members'—let alone its leaders'—names.

Despite Trocmé's pleas and those of his wife, Le Forestier set out for Le Puy. On the way, he picked up two young Maquisards who needed a lift. Before he allowed them to enter the car, he asked if they had weapons with them—he would not take armed

Maquisards anywhere. They said they had none. The three arrived in Le Puy safely, and the two young men went to the nearby terrace of a café for a drink while Le Forestier went to the prefecture of Le Puy to plead for the release of the two prisoners. The prefect refused to help. (Actually it turned out that the whole project was unnecessary—the prisoners were released before the Gestapo could deport them.)

Le Forestier left the prefecture and returned to his car. Trocmé says that he was ambushed by the Gestapo, who had been watching for the owner of the car to return. Others say that the Gestapo were examining his car when he arrived and he imprudently walked right up to them and identified himself as its owner. In any case, he was beaten severely, thrown to the ground, and lost two teeth. All of this happened before the eyes of the two Maquisards, who were sipping drinks on the terrace of the nearby café.

What had happened was this: the Gestapo had been alerted because a bank robbery had happened at the time Le Forestier and the Maquisards were entering Le Puy. It was one of the many crimes being committed during these disorderly, violent times. The Gestapo found a civilian car parked in front of the prefecture, and under the cushions of the rear seat they found a loaded revolver, which the two Maquisards had left hidden there without telling Le Forestier.

Le Forestier was taken before a German court-martial and accused of a plot against the German Army. The prosecuting attorney was Colonel Metzger, the SS officer in charge of the Tartar Legion, which was now in Le Puy waiting like hungry wolves to be released so that they could strike down their prey. The prey of the Tartar Legion was dangerous civilians, hostages, and troublesome villages. The main evidence against Le Forestier was the loaded revolver that the Gestapo had found in his car.

Apparently Le Forestier made a magnificent defense. He asserted with all the warmth of his youthful personality that he did not know of the existence of the revolver in his automobile. And then he gave the court a moving account of his belief in Christian nonviolence. Here is how Trocmé summarized his words: "We in Le Chambon resist unjust laws, we hide Jews, and we disobey your orders, but we do this in the name of the Gospel of Jesus Christ."

The head of the tribunal was the commandant of the German Army post in Le Puy, Major Schmehling. He was not a professional soldier, had taught at a secondary school in southern Germany, and was in the army reserve. He was a devout Catholic. Le Forestier's testimony convinced him completely that the young doctor was innocent of any plot against the German Army.

But the prosecuting attorney, Colonel Metzger (whose name means "butcher" in German), demanded that some punishment be meted out to Le Forestier if he was not to be condemned to death. After all, the evidence for his "possessing" an illegal weapon was conclusive. Major Schmehling devised a solution to the problem: he had Le Forestier acquitted as far as the charge of having committed a capital offense was concerned, but he asked the doctor to volunteer to go to Germany to tend civilians who had been wounded in the bombings. In Germany there was a great need for doctors now that Allied bombing was becoming more intense than ever.

Although he was not legally bound to do so, Le Forestier signed a document pledging that he would go to Germany to help the wounded. He did this out of gratitude to Schmehling for having saved his life.

His wife Danielle came to see him before he left for Germany. Their farewell was not without hope. After all, he had been acquitted of the capital charge.

Soon after the liberation of southern France, Danielle started a search for her missing husband. In the course of the search she learned that there had been a massacre of prisoners near Lyons, and that her husband might have been one of the victims. She went to the prefecture of the department in which Lyons is situated and told her story. They informed her that there had been one hundred and twenty victims of the massacre, but all the remains they had were in one hundred and twenty little sacks containing the personal possessions of the victims. One after another, she examined the sacks and found nothing. She turned to the last sack, the very last. In it she found a button on which the name of her husband's tailor in Montpellier appeared, and a piece of what she recognized to be her husband's underclothing. These are the items Trocmé gives in his notes; another source says that she found only the buckle of his belt. In any case, his death was confirmed.

His wife tracked down some of his fellow prisoners in the Lyons prison. One told her that her husband could have escaped on the trip from Le Puy to Lyons; some prisoners had succeeded in doing so, and they had invited him to come with them. But he had refused, saying that he had given his word of honor to go to Germany to help the wounded. Another told her that he was always as happy as a boy; he sang hymns, read the Bible to those who wished to hear it, and by his example and his words he gave courage to everybody. His fellow prisoners loved him.

This was her husband to the life; now she knew that he was, in fact, dead.

3.

The story became fuller long after the Liberation in 1944. In the 1960s, Magda and André Trocmé found themselves in Munich, where André was lecturing on nonviolence for the International Fellowship of Reconciliation. Knowing Major Schmehling, who had been a prisoner of war immediately after the Liberation, but who had come back to Le Puy a few years later to receive in a formal, warmhearted ceremony the gratitude of the people of the region for all his deeds of kindness, the Trocmés went to visit him one afternoon.

He lived in a house still partially gutted by bombs. They rang his bell, and after a slight hesitation, he recognized them. "*Ach!*

Pastor Trocmé!" he said. "Naturally! Come in!"

After a while, Trocmé said, "I am here to ask you two questions, Herr Schmehling. The first is: You knew that Le Chambon was a nest of resistance; you knew we had Jews there, and the Maquis nearby. It is true that your German police did us harm, but why did you not send a punitive expedition to destroy the village in those last months? Surely you were doing this elsewhere in France, and in places near Le Chambon. . . ."

"Monsieur Trocmé," he answered, "it is difficult to answer that question. You know that we had in the department of Haute-Loire the Tartar Legion under SS Colonel Metzger." Trocmé knew of the man who had been the prosecuting attorney in Le Forestier's trial, and who had been executed for war crimes after the Liberation. "Well," Schmehling went on, "Colonel Metzger was a hard one, and he kept insisting that we move in on Le Chambon. But I kept telling him to wait. At his trial I had heard the words of Dr. Le Forestier, who was a Christian and who had explained to me very clearly why you were all disobeying our orders in Le Chambon. I believed that your doctor was sincere. I am a good Catholic, you understand, and I can grasp these things."

Schmehling went on, "I told Metzger that this kind of resistance had nothing to do with violence, nothing to do with anything we could destroy with violence. With all my personal and military power I opposed sending his legion into Le Chambon."

Then Trocmé asked his second question: "But if you believed all this, why were you not able to save Le Forestier?"

Schmehling, now a retired schoolmaster whose life after the war had had nothing to do with the army, answered, "In order to save Le Forestier, I had to fight Metzger and other officers, and it cost me much, professionally and personally, to do so. Nonetheless, I told them that I was certain that this man was not dangerous, and I commuted his punishment from death to acquittal if he worked in Germany. It was all I could do."

"And then?" Trocmé pressed.

"And then," said Schmehling, with his fists clenched and tears in his eyes, "and then those *Schweinehunde* in the Gestapo in Le Puy must have sent a message to their colleagues in Lyons to take Le Forestier out of the group that was going to Germany to work. Anyway, after he was taken out of the group, he was put in the Montluc Fortress at Lyons. There he was killed. *Voilà.*"

"*Voilà,*" Trocmé found himself repeating in French.

"And this is why," Schmehling went on, "I come awake at night, even now, after my nightmares. In my dreams and right now I see that beautiful, beautiful young woman and her two children who came to beg me to help Le Forestier. And I see them taking leave of that man. And she had confidence in me—yes, confidence that he would return. What must she think of me?"

The old teacher sat before the Trocmés with tears in his eyes, but Trocmé could not lie to him. He said to Schmehling, "She is not yet able to pardon you." And after a little while the Trocmés left him.

4.

Schmehling did not know the details of the killing, but Trocmé and others have confirmed them. After the Gestapo took Le Forestier out of the group going to Germany and put him in the prison of the Montluc Fortress, some of the police attached to the Gestapo, after a night of drinking, went down to the cells of the fortress early in the morning and took a large number of people —men, women, and children—out. Among them was Le Forestier. They brought them to an abandoned farm, where they killed them with machine-gun fire, piled them up, sprinkled them with gasoline, and set fire to them. Neighbors heard the screams of those the bullets had failed to kill. The date was August 20, 1944, a few days before de Gaulle's triumphant arrival in Paris.

And so Roger Le Forestier died. In his notes, from which most of this account of Le Forestier has been drawn, Trocmé calls him the *puro folle* (purehearted fool) of Le Chambon. In those notes he goes on to describe him as the least cunning *(combinard)* and most idealistic person in the village. He was springtime, a joyous child, despite his great knowledge and skill in medicine, and despite his profound commitment to nonviolence and to the imitation of Christ.

It was fitting that he was the one who strengthened Schmehling to keep the SS from destroying Le Chambon. He was more than a symbol of spontaneously creative living; he was its incarnation.

There is a kinship between him and another *puro folle,* Daniel Trocmé, who, in the words of Magda, "worked like a madman" for the children. The words *fool* and *madman* suggest a certain impracticality, a certain lack of ordinary, commonsensical self-interest, a certain amateurishness that they shared. Daniel Trocmé was no joyous child; he was duty working its heart out for the weakest of the weak, the children. But both of these young men celebrated and defended life, and their deaths cause us to celebrate life by making us hate murder.

11

The Astonishing Weeks

1.

Between June 1944, when the Allies landed in Normandy, and September 1944, when Le Chambon was liberated by French troops, life was packed with absurdity and death. During this time, a faithful adherent of nonviolence, Le Forestier, was arrested for possessing arms that were not his, and was killed by German police when he was supposed to be on his way to helping wounded Germans. And during that last summer of the Occupation, two adolescents whose parents had done much to save the lives of refugee children died, and died not at the hands of enemy soldiers or police, but by little accidents.

In 1942, Albert Camus spent about a year in the region of Le Chambon. He lived in an old granite house named Le Panelier, within easy walking distance from the center of Le

Chambon, and in that house he began writing his novel *The Plague.* In some ways the story of Le Chambon is a companion piece to this novel. In both stories there is a leader totally dedicated to saving lives; in both stories not only courage but *realizing,* comprehending the horror of dying, plays a central role in the leader's mind and in the minds of his closest associates. Trocmé saw more clearly than anybody else in Le Chambon the danger to the refugees, and he saw with equal clarity that they must be saved by nonviolent means if they were to be saved at all. In *The Plague,* Dr. Rieux (with the help of his friend Tarrou) clearly saw the dangers of the plague, and saw with equal clarity how to organize the city of Oran to fight them. Both men knew that the plague of mankind is man's desire to kill, or, more usually, man's willingness to allow killing to happen without resisting it. And in both stories the plague disappears as terribly as it came, pointlessly killing a person here and a person there, and then subsiding.

Of course, as companion pieces should be, the two stories are different. One is fiction, and the other is true; one is mainly the story of a committed atheist, and the other is mainly the story of a committed Christian. But the two stories are about the tenacious saving of lives in a dangerous time,and they are about what Rieux's closest friend, Tarrou, describes as the "good man, the man who infects hardly anyone" with the microbe of hatred and indifference. And they are about how one may fight death on one front and be struck by death on another. In the novel, the dearest friend of the doctor dies a few days before the plague is past, and so does the doctor's wife, even though she has long left plague-stricken Oran. In the story of Le Chambon, the eldest, closest son of André Trocmé, Jean-Pierre, died less than a month before the liberation of Le Chambon. In both stories, death comes absurdly into the intimate life of the "good man."

2.

When, shortly after the landing in Normandy, Trocmé came back to Le Chambon, it looked as if the village were another Célisse: it seemed to have been committing suicide in his absence. True, the Germans, sensing the coming of defeat, had escalated violence in the area by shooting at unidentified moving civilian vehicles, by killing hostages, and by threatening the whole Haute-Loire department with Metzger's Tartar Legion, those Asiatic-Russian prisoners who had been captured on the Russian Front by the Germans, and who had been dressed up in SS uniforms and trained to murder civilians without mercy. But whatever the provocation—and the civil war that was boiling up between Frenchmen was a part of that provocation—it seemed at first glance that Le Chambon had in his absence committed itself to living and dying by the sword. Many of the young men of the Cévenol School had joined the Maquis and were helping them use the new plastic explosive that could stick on railroad tracks or bridges and blow them up. And in the mountains of Haute-Loire, young men and women, working with hardened Maquisards or new bandwagon "patriots," were ambushing German troops.

But a few minutes' conversation with Magda in the presbytery showed him how misleading appearances were. Le Chambon was not another Célisse. Magda and her equally aggressive friend Simone Mairesse, whom she had known in their first parish in Maubeuge, had turned Le Chambon into one of the most important underground railroad stations in the south of France. The two women had been welcoming ever-increasing numbers of refugees into the presbytery and into the various types of shelters in Le Chambon. The houses of Le Chambon were always full, and Cimade teams were busier than ever making room in Le Chambon for new refugees.

Mildred Theis, in her quiet, endlessly generous way, had helped keep the Cévenol School a hospitable place for refugees, and so had the remarkable faculty of that school. In the village itself, the women who led the *pensions* and the private homes had been taking on more and more refugees with fewer and fewer funds for feeding them; and in the countryside, the Darbystes and others were helping the women of the town to "maintain a spirit of peace in the parish and the village," as Theis described it all more than thirty years later. And it was the same sort of peace that Le Chambon had had with Trocmé and Theis there—not the peace of retreat, but the peace the Chambonnais felt in living conscientiously, in taking, as Camus puts it in *The Plague,* "the victims' side, so as to reduce the damage done." Happy André Trocmé! To have such friends after poor Célisse!

3.

But death, like the biological microbe of inguinal fever in Camus's Oran, was in the air of Le Chambon. It was as if during those "astonishing weeks," as Trocmé later called them in his notes, man's murderous intentions infected the atmosphere of Le Chambon and became depersonalized but stayed murderous. People died because of something in the atmosphere, something we can dismiss with the word *accident,* but something that must be looked at more closely if we would understand both Le Chambon and certain of the people of Le Chambon as they were during those weeks.

One floor of the boardinghouse of Madame Barraud had been requisitioned by the Maquis for young Maquisards who had gone into the organization to avoid forced labor in Germany but who were now unable to get back to their homes to fight with Maquisards in their own regions because trains were not moving regularly, bridges were blown up, and the dangers in the countryside were too great.

Madame Barraud had four children, one of whom, Madeleine ("Manou"), was eighteen years old. Manou had tiny, efficient hands that seemed able to do anything speedily and perfectly. She was a close friend of Jean-Pierre Trocmé, though he was four years her junior. Madame Barraud did not approve of their friendship. It seemed to her that Jean-Pierre was even more violent than his father, the pastor of Le Chambon; there was a driving force within the two Trocmés that frightened some of the people who knew them intimately. Madame Barraud felt that just as André Trocmé had led the village into harm's way with his powerful passions and firm will, so Jean-Pierre might lead her daughter, Madeleine, into danger. Madame Barraud, like others in Le Chambon, had followed André Trocmé down a dangerous path because she knew that it was her path too, her only way to go, a path of helpful, unsentimental love. But she was far from sure about Jean-Pierre, whose anger sometimes seemed unbridled to her. He was not a good Boy Scout, an important organization in the village, and he disliked going to temple. Manou, on the other hand, was devoted to the Scouts, especially the *louveteaux* (Cub Scouts), and was deeply pious. Manou had very strong feelings against the Germans and had more than once been dangerously outspoken against French girls who had anything to do with German soldiers. The girl might be led into a drastic act by the wild Jean-Pierre.

But one day Madame Barraud discovered what sorts of things the two adolescents did together. One of their most frequent activities was that of bringing food and wood to the old people in the village who could not get about. They took a special pleasure in finding coffee for them. And they were only comrades, not lovers—Jean-Pierre was too young, though he looked a bit older than his fourteen years, and Manou loved another young man, who was away fighting with the Maquis. The two shared one passion, aside from their passionate dislike for the German occupants of France: a passion to help. The French phrase *toujours prête* is not often used by Chambonnais. It is for them an abbreviation of the phrase *toujours prête à servir* (always ready to help). Dr. Riou, who was one of the sponsors of the Scouts in the area, said that of all the Scouts, Manou was the only one who was *always* ready to help, with those golden hands of hers and that slender, ardent, dark-eyed face. And everybody in Le Chambon who knew her at all well used this phrase, the highest accolade the Chambonnais can give, about her. Jean-Pierre was also *toujours prêt* (always ready), but more erratically, more impulsively.

In early July of that last summer of the Occupation, on a splendid day just before the golden flowers of the *genêts* started dying away, Madame Barraud took some children from outside Le Chambon for a long walk around the village and into the countryside. She had invited her daughter to come with her, but Manou had already committed herself to taking care of some Cub Scouts.

Manou finished her duties early and returned to the Barraud boardinghouse in the center of the village. Since it was July and there was no school, and since her high-spirited health demanded activity, she joined a young man who was a boarder in the house, but not a Maquisard, on a tour of some of the empty rooms on the floor of the boardinghouse that had been requisitioned by the Maquis. Some of the Maquisards were away, and the two young people wanted to see their sometimes dangerous but always interesting possessions, which they would carelessly leave about in their rooms.

Her young companion found a revolver that a Maquisard had left in the drawer of his night table. He picked it up, casually checked to see if there were bullets in the gun, and then playfully pointed the gun at Manou to frighten her. But there was a bullet in the chamber of the revolver, and when he pulled the trigger, he shot Manou, hitting an artery. She died almost at once.

When Madame Barraud returned to her boardinghouse, someone ran up to her and cried, "Manou is wounded! Manou is wounded!" She found her daughter dead.

When she saw that there was nothing she could do for her daughter, she immediately asked, "But where is Jean?" Jean was the boy who had shot her. Somebody had seen him run from the boardinghouse, his eyes wild with grief and self-hatred. Madame Barraud pursued him, took him back into the house, embraced and comforted him, and told him, "Jean, it is not your fault. It is not your fault. It is the fault of the war. We are all a bit mad. It is not your fault."

When the temple funeral service was over, Madame Barraud looked around her, trying to find Jean. He was not in the temple. She knew that he could be in only one place, if he had not killed himself, and so she rushed to the fresh grave of her daughter. There he was, lying on the soil above her coffin, prostrate, sobbing. She took him in her arms and soothed him. Miss Maber, who knew Jean, told me about her efforts to console him. And then she added with her eyes shining and her body shaking off its customary English poise, "She was absolutely wonderful. I cannot say enough about how wonderful I think Madame Barraud was. She is a glorious woman." The plague had struck, but the little Alsatian had remained "always ready, always ready to help."

4.

About a month later, on August 13, 1944, the pastor and his wife went to visit two Swiss friends in Le Chambon who had just been attacking each other verbally with such bitterness that their friendship was in danger. The Trocmés went, as experts in such turbulence, to help reconcile them, both to each other and to a moderate amount of friendly quarreling. During the latter part of the visit, Magda kept pressing André to leave early, but he wanted to stay so that they could eat the "cake of peace" with their Swiss friends. As they approached the presbytery, Magda said, "If only I do not arrive too late!" She does not know now, and believes that she never knew, why she uttered these words, but she was being drawn to the presbytery. Trocmé left her at the "poetic gate" and was walking back up the Rue de la Grande

Fontaine to go to the temple when he heard one of his children calling after him, "Papa, Papa! Come quickly! Jean-Pierre has hanged himself!"

When he found him stretched on the floor of the bathroom, the boy was still warm and smiling mildly, as his beautifully proportioned, bronzed young body lay there, with only underdrawers on, dead. "I have just taken him down," Magda said, in a voice as flat as death. And she went on, "Oh, but he was heavy! He tied the rope up there on a pipe. I don't know how I got him down. He is warm! Go quickly! Get the doctor. Maybe he is still alive."

But he was not breathing. For the Trocmés, who did not know about mouth-to-mouth resuscitation, he was dead. When Dr. Riou arrived, the boy was cold.

The evening before, Jean-Pierre had accompanied his parents to a recital by the distinguished actor Deschamps of François Villon's poem "The Ballad of the Hanged." The boy knew the poem by heart, and, mesmerized by Deschamps's stiffening and swaying body and by his affecting rendition of the lines against an austerely simple background in the annex of the temple, he had begun to recite the verses along with Deschamps and had moved up the center aisle of the annex hall like a sleepwalker. It was the carefully considered opinion of Dr. Riou that at about four o'clock on the following afternoon, when Jean-Pierre was alone in the presbytery and depressed about a difficult Greek translation he was working on, and about not finding his friends at the Lignon, where they had agreed to swim together, he decided to pass the time by imitating Deschamps's recital. He wanted to see how he as a hanged man would look swaying at the end of a rope, so he went to the bathroom, knotted a rope around the pipe, made a hangman's noose with the other end, found a stool, stood on the stool before the bathroom mirror, and put the noose around his neck; then, while he was perhaps swaying and reciting the lines of Villon's poem, his foot slipped and the noose pressed a nerve, rendering him unconscious before he could recover his footing. Unconscious, he was strangled. This was Dr.

Riou's account. We shall never know with absolute certainty what happened in the lonely presbytery.

Of his four children, Jean-Pierre had been the closest to André Trocmé. His powerful imagination, which had, in all likelihood, brought about his death, his immense sensitivity—he was an accomplished pianist and a vigorous young poet—brought him close to his restless father. One day he was playing the piano in the dining room, and his father walked up behind him, put his hands upon the boy's wide shoulders, and said, "Jean-Pierre, you are my oldest son. Someday you will take my place, and you will continue what I have been doing, and you will do it better than I have done it, at least in certain areas. I can count on you, can't I?" The boy did not answer and simply resumed his playing. But his big shock of stiff hair was at the level of his father's lips, and his father felt upon his lips through that hair the boy's love for him and his confidence in himself. They both knew what Madame Barraud and most others did not know, that Jean-Pierre was deeply pious, though he did not go to the temple. Since the age of thirteen he had gone down to the lower village to pray with the old people when his father was away or busy. The old people called him "our friend." His passions, moving in so many directions, often made him absentminded, so that when people spoke to him during one of his imaginative flights, he turned his eyes upon them with surprise and said, "What?" That "*Quoi?*" was the typical Jean-Pierre response.

Seeing him dead, Trocmé remembered the feeling he had had at seeing his mother dead by the side of the road when he was nine years old. There was that same contradictory feeling of sheer emptiness filled with pain, nothingness overwhelming in the fullness of its power. He and the rest of the Trocmés had rationalized the death of his mother. In 1914, his father used to say, "Fortunately, your German mother is dead. She would not have been able to bear being rent in two by our two countries and our two families in this terrible war." But there was no way to understand, to rationalize the death of Jean-Pierre. And not even his faith in

God could reconcile André Trocmé to that death. Soon after his son's death, he marched into the nearby woods in his driving, pain-filled way, in the hope that God would give him a vision that would make that death something other than a horror to him. Pounding through the woods, he screamed at the top of his voice, "Jean-Pierre! Jean-Pierre!" Maybe the boy would answer from Heaven. But there was utter silence, except for the blankly meaningless signs of an indifferent nature. And inside him there was only the paradoxical feeling of horror-surrounded solitude, pain-filled nothingness.

In the notes he wrote in his sixties, he says that he suddenly understood that human life is thrown into a world dominated by absurd and chaotic accidents instead of a world providentially ordered by a loving God. In his notes he wrote: "I had, without knowing it, joined Sartre and Camus, who were unknown at this time." Sartre's idea that each person is to himself a dark, useless hole in a full, pointless world, and Camus's idea that all our joy and all our valor come and go because of absurd circumstances and not because of any rationality or love in the universe, were at the center of Trocmé's grieving and surprised mind.

Never before, except perhaps at the age of ten when his mother died, had he fully realized how precious the life of another person is; but never again would he believe that God protects that precious life. Never again could he pray to a Protector-God. From now on, God and Jesus were to him powerless, suffering, limited. God was still the Father, but He was as powerless as Trocmé the father was. God could only join us in our grief, not save us from it.

There is an image in Trocmé's autobiographical notes that is, perhaps, derived from that horrific walk in the woods when he screamed for help and received none. More than thirty years after the event, he wrote:

> Even today I carry a death within myself, the death of my son, and I am like a decapitated pine. Pine trees do not regenerate their

tops. They stay twisted, crippled. They grow in thickness, perhaps, and that is what I am doing. I think less now and I throw myself into action more. Under a scar that the years have thickened little by little, there bleeds in the very depths of my being an incurable wound: an awareness of nothingness, a total resignation before nothingness, nothingness toward which I am going, and toward which all of those around me are going too.

These are not the words of a person permeated to the very foundation of his being with a peace-filled faith in God. He wrote in his notes: "Rarely do I *possess* peace. I have received it from God as a gift, but at the bottom of myself all problems keep posing themselves anew for me again and again." He was a turbulent man who looked like a great pine to others, but felt like a volcano ready to explode. The peace he felt in loving God and people crowned his life; it did not pervade it. His grief over the death of Jean-Pierre tore off from atop his mind the peace that had been laid upon him, and revealed himself to himself as a churning, molten mass always thrusting itself restlessly outward.

Magda's reaction to the death of her eldest son was different because Magda was different. After the coffin had been carried to the cemetery behind the temple by the remaining members of the family and some of Jean-Pierre's friends in the Maquis (one of them reported later that Jean-Pierre's body had gotten very heavy suddenly, as if he were reluctant to disappear from among them), and after the coffin was covered with soil, flowers were put upon the tomb, flowers tied together with multicolored ribbons by the hands of children. During the burial ceremony, Magda kept repeating, "My poor little one. My poor little one. If I had been there—I should have been there. I had a feeling something was wrong. Gone like that, all alone, all alone, without a good-bye." And she cried.

The next morning, they went back to the cemetery, and Magda bent and started to take the ribbons off the bouquets of flowers. Her husband asked her, "Why are you doing that?"

She answered, "I'm taking these ribbons for the refugee children who have replaced Daniel Trocmé's refugees at the House of the Rocks."

After writing down this story, Trocmé wrote: "Then I understood that she was coming back to health faster than I was. She had not lost her awareness of the external world. I had." And he went on: "She was more valiant than I was in the moment of greatest grief." Her responses were, as always, concerned with the well-being of human beings. She cried, but her husband could not. She responded simply to people; her husband responded also to his own wild soul and to his God. The morning after the interment, she took the ribbons off the flowers so that the children, who were alive and in need, could enjoy them. With all his religious faith—changed, but remaining—all he could do was suffer and find the same suffering in his God.

Magda Trocmé was still Magda Grilli, the same person André Trocmé had met in New York, a person who covered and uncovered the needful bodies of people. She could suffer as deeply as her husband, but always her suffering swiftly issued in humane action. This is what Trocmé meant when he described her as "more valiant than I was."

Once after the war when she was lecturing in America and Europe about those years in Le Chambon in order to explain nonviolence and to raise much-needed money for the Cévenol School, she brought with her a pair of trousers that had been worn by Jean-Pierre during the war years. In the course of the lecture she showed her parish friends those trousers with their patches of different colors and different materials in order to make vivid the poverty of the village and the school during those years. A few years after the death of André Trocmé in 1971, she decided to destroy those trousers because she felt that she would soon die, and she did not want the trousers to be handled by somebody who did not understand what they meant.

But when she told me this story, I was shocked; how could she carry the garment of her dead child into a public room and hold it up and speak about it? I never asked her this question, and I never shall. When I came to know her better, the answer became plain. She is a doer as much as she is a feeling person. I did not have to ask her if every time she touched that garment she suffered. Her nature is as passionate as her husband's was, and she loved that boy who had tried to get away from her to prevent the Gestapo agent from beating the Jewish refugee child with his own phylacteries; she cherished that son who had been the first to give a gift of precious chocolate to Monsieur Stekler as he sat arrested in the Vichy bus. But she is also somebody who is "always ready, always ready to help."

The death of Jean-Pierre expunged whatever religious feeling she may have come to have over the years. During the "awakening" at Sin-le-Noble that saved Célisse temporarily, her criticisms of Christianity had softened considerably, and ever since had been friendly criticisms. But now she turned her back on all religion, and on her husband as pastor, so that their marriage for a while was very painful, and later her criticisms of religion went back to their old severity. But their personal and physical love for each other kept the marriage thriving, and their shared eagerness to diminish suffering and killing in the world gave that marriage meaning for them both.

5.

On the day of Jean-Pierre's burial, August 15, 1944, the BBC announced the landing of the Allies in the south of France. From then on, the Germans were caught between two great military forces—the triumphant Allies in the north and the equally aggressive troops in the south. The battle for France was in its last weeks. But for the most part, the Germans kept their discipline in their retreat from southern France. One hundred and twenty of them were seized on the northerly road to Saint-Étienne; the Maquis, too, were well organized. But in the streets of Saint-Étienne, German prisoners were attacked, and two of them were lynched; in the neighboring department of the Ardèche, according to Trocmé, forty-five German prisoners were found in a well, massacred.

But in Le Chambon, Trocmé did much to keep such massacres from happening. The one hundred and twenty German prisoners of the Maquis were put in the artificial castle on the Lignon called the Castle of Mars. Among them was their chief, Major Schmehling. Trocmé (partly to set an example for the region) went to the castle to preach to them, and Schmehling, though he was a Catholic, did all he could to get his soldiers to go to the Protestant services. And many of the German soldiers went.

Trocmé's fellow Frenchmen, often members of the Maquis, objected to his sermons against war, sermons based upon the Ten Commandments and upon the belief that Christ had shown us that we must forgive sins instead of killing the sinner. They told him that nonviolence might be true in theory, but in practice, with people like "those guys there in the castle," only force counts—and they reminded him of the recent German massacres and of the gas chambers in the death camps.

When he gave the identical sermon, in German, to the prisoners in the afternoon (he gave his sermons in the temple in the morning), the Germans listened politely and told him afterward that their own brave soldiers were ridding Europe of the Red Plague, Russian Communism. When he mentioned the massacres and the gas chambers, they raised their shoulders unbelievingly and said, "Filthy, lying war propaganda."

Both sides were utterly convinced of their own innocence and of the guilt of their adversaries, which justified any killing of "those guys there." The end of the war in France (though some of the Germans still thought that Hitler would save them with a new *Blitzkrieg*) had made no difference; people were bent on killing each other because of their own innocence and the others' guilt.

The last refugee train to arrive in Le Chambon was the "phantom train." Among the various trains of political refugees the

Germans were bringing back to Germany, one was sidetracked, taken by the Maquis, and brought to Le Chambon with its fifty amazed and joyous escapees.

In September, General Jean de Lattre de Tassigny's troops liberated Le Chambon, and Jacques Trocmé remembers all the impressive vehicles and all the black soldiers that passed along the south side of the square, with the people cheering and throwing flowers and kisses at soldiers who seemed only mildly, perfunctorily appreciative of gestures they had been witnessing for weeks.

Soon the refugees left, the children's houses closed, and so did the funded houses like the Cimade's Flowery Hill. The school was reduced by half. Even the Maquis left, to join the Army of the Rhine for the final battles with Germany. Le Chambon seemed to have returned to the sad and dying life Trocmé had found when his family first arrived there a decade before.

Both the villagers and the farmers felt a sense of slackening after four years of being drawn along by events, by their leaders, and by their consciences. The heads of boardinghouses, like the Marions, Madame Eyraud, and Madame Barraud, even then began to see that during those four years they had lived more vivaciously than ever before, and they felt lucky to have been able to do something of use. But their feelings were not only directed at the past. They felt "relief and also joy," as the Ernest Chazots put it, now that those last months were over. And the little Alsatian, Madame Barraud, says that they trusted themselves and each other more than ever.

Trocmé, at forty-three years of age, felt tired, "like a fighter," as he puts it in his notes, "exhausted by the battle." He felt that he had lived through the harshest and most useful years of his life. But his spirits rose when he thought of the people of Le Chambon, especially during those last few months. He thought of their common sense, their humor, and above all their courage in disobeying desperate enemies who felt now that they had nothing to lose by more killing.

But now those years were past, and Trocmé had to build a new life, just as Le Chambon had to build a new life. It was fall, and the village seemed brown and gray and empty, with two-thirds of its houses shuttered, and those who were left were beginning to settle into the old habits of hibernating, of looking forward drearily to the summer season when the tourists would awaken them to three months of frantic moneymaking after nine months of deadly winter.

What lay ahead for Trocmé and the village was the rejuvenation of the Cévenol School, by way of a trip Trocmé would take to America, where he would find friends for the school, people who would give money to it and work for it. The village would come alive again, all year round, unlike her quiet companion of The Mountain, Le Mazet.

6.

But they would never be the same, and the one who had changed the most was André Trocmé. The people who remained close to him during the war years did not see any sharp change in him, but those who had been away from Le Chambon for a long while saw it. They had always known a strong-willed man, with powerful passions and the courage to displease the smug and the powerful. But now they found a man who had become *weightier,* more authoritarian than ever, more accustomed to being obeyed. He had developed a commanding manner that had been not only appropriate during the war, but necessary, especially in moments of crisis like the raid of the Vichy buses or during an argument between the Maquis and the villagers or peasants. He had grown accustomed to going for months on end without convening his presbyterial council for advice, and the word *I* occurred more and more often in his sermons. The man who could work intimately with the people of Maubeuge and Sin-le-Noble and Le Chambon so as to have deep, one-to-one relationships with the members of his parish had become accustomed to power and had lost much of his capacity to do things *with* the humblest members of his parish. As one Chambonnais put it, "He led the parish with a rod."

And this change was involved with another change in him: he had become interested in issues larger than the spiritual leadership of a country parish. More and more, he was thinking about nonviolence on an international scale, and after the war, especially after the Cévenol School was launched again, he lectured frequently for the Fellowship of Reconciliation across the face of Europe and in America. The conversations he had had with his *responsables,* those who had helped him guide the parish into nonviolent action through Bible discussions, had accustomed him to conversations at a level of intellectualism higher than that of the average Chambonnais. One man who returned to Le Chambon after the war was amazed to see how far beyond both the interests and the capacity of the Chambonnais his talks could be.

Burns Chalmers, who had helped him make the decision to make Le Chambon a village of refuge, had seen him even at the beginning of the Occupation as a "major figure," as somebody with too much imagination and too much compassion to become a bureaucrat or a servant of bureaucrats. He had seen him as a man capable of taking large steps through the "great openings" that the Quaker George Fox had seen for those who would work for love. Chalmers had been right—both with regard to the Trocmé of the Occupation and with regard to the Trocmé after it. By the time the Occupation had ended, Trocmé had become a weightier man than a little mountain parish could bear, and in a few years he left Le Chambon to become the European secretary of the Fellowship of Reconciliation, the world's most effective organization dedicated to international nonviolence. Still later, he became pastor of Saint-Gervais, one of the oldest parishes in Geneva, that most international of Western cities. He held this post until his retirement. He died in 1971, a year after he retired from the ministry.

André Trocmé was too energetic, too surprisingly creative to be categorized neatly. But there was one description of himself that even he accepted, which was that of *un violent vaincu par Dieu* (a violent man conquered by God). Those who knew him best, including Édouard Theis, who is a profoundly nonviolent man, accepted this as the most apt description of him. There was a tension in him between anger and love that took different forms in different parts of his life. In his childhood it took the form of anguish at his mother's death; in his youth it took the form of the union at Saint-Quentin and of his tense admiration for Kindler, the German soldier who taught him nonviolence. In his early parishes it took the form of a passionate mysticism controlled by Magda Grilli Trocmé's compassionate common sense, and issuing in the "awakening" of Sin-le-Noble and the fragile redemption of Célisse. The form it took in Le Chambon is to be seen in a certain passage he wrote in his notes about the summer of 1939, when Hitler was approaching the peak of his audacity and power. Trocmé had just described Nazism as a way of unleashing the "diabolical" forces in mankind by using violence and lies. Then he wrote about his thoughts at that time:

> . . . should I not make use of my knowledge of German to slip into Hitler's entourage and assassinate him before it is too late, before he plunges the world into a catastrophe without limits? It is because I feared separating myself from Jesus Christ, who refused to use arms to prevent the crime that was being prepared for him, and because of a kind of stubborn perseverance in the growing darkness that I stayed in place; it was also because my ministry in Le Chambon was becoming more and more interesting.

Only his complex love for Jesus, for commitment itself, and for the fresh, the "interesting," could overcome this anger in him against Hitler, who was doing such harm to God's precious human lives.

Those words—like the rest of his notes and all of his life—reveal a man of such intensity of awareness that he seemed to be a microcosm of all humanity, seemed to contain within the limits of his own skin both the world's destructive forces and the world's creative forces. He realized, or, to use Camus's word, he comprehended that struggle because he reenacted the struggle within his own mind.

He once described Célisse as "nature tamed, a sort of hurricane made useful for turning mills." The image applies to Trocmé himself. His love for human life, a love deepened by his love for the words and deeds of Jesus, was so stubborn that it tamed his immense capacity for anger and made his energies "useful for turning mills."

But even if these images are misleading or wrong, one point is clear: he was not one of those vapid "good people" whom antimoralists over the centuries have despised. Whatever moral power he had was in total opposition to pusillanimous conformity. He believed that "decent" people who stay inactive out of cowardice or indifference when around them human beings are being humiliated and destroyed are the most dangerous people in the world. His nonviolence was not passive or saccharine, but an almost brutal force for awakening human beings. He earned the description that the national leadership of his church had made of him: "that dangerous, difficult Trocmé."

In the last pages of his autobiographical notes, Trocmé wrote: "A curse on him who begins in gentleness. He shall finish in insipidity and cowardice, and shall never set foot in the great liberating current of Christianity." Kindness, in order to be valuable, had to be achieved through a struggle with the forces for humiliation and death that are part of mankind. The modern antimoralist Friedrich Nietzsche once wrote: "You must have chaos in you to give birth to a dancing star." The chaos of aggression and love in André Trocmé helped give birth to the dancing star named Le Chambon.

PART FIVE

The Ethics of Life and Death

12

How Goodness Happened Here

1.

One evening, Roger Darcissac, now the official historian of Le Chambon, saw a television program that pictured the liquid-crystal Lignon racing through Le Chambon, the thick pine forests in and around the village, and the vast mountains and extinct volcanoes surrounding it. At the end of the program, the announcer mentioned the wartime activities in the commune, and then ended with the sentence: "You must not forget Le Chambon." A few weeks after the program, Darcissac and I were eating in the sleepy village of Le Mazet. Suddenly, in his impetuous way, he seized my arm and said, "You know, they should not have spent so much time on the scenery. The reason people should remember Le Chambon lies in the people there. Ah! During those years they were completely wonderful *(merveilleux d'un bout à l'autre).*"

In his autobiographical notes, André Trocmé said the same sort of thing about the Chambonnais. He wrote: "How could the Nazis ever get to the end of the resources of such a people?" Like Darcissac, he was passing an ethical judgment upon the facts in the story of Le Chambon during the war years.

An unrepentant Nazi might study the same facts about that "nest of Jews" in Huguenot country and might judge the Chambonnais differently. He might see their actions as the actions of a nest of vipers, a group of "Jew-lovers" incapable of patriotism. Could Trocmé and Darcissac explain and defend their praise successfully before a court that they and the Nazi would both accept?

I believe not, because in matters of ethical praise and dispraise there are no such courts before whom cases may be pleaded "successfully." Ethical judgments, though they often use words that look like words used in law courts (words like "innocent," for example), are deeply different from the judgments laid down in such courts.

Criminal law, for example, moves and lives in public institutions; life-and-death ethics (which is the area in ethics closest to criminal law) moves and lives in individuals. Legislatures frame laws regarding criminality; police bring force to those laws; courts make judgments upon particular cases by applying and interpreting the laws; and prisons administer some of the punishments laid down in the courts.

None of this happens with ethical "judgments." The origins, enforcement, application, and punishment involved in moral "laws" are idiosyncratic, personal, and are subjects of endless disputes. Who created those laws? Where is the force that makes the laws effective? Who is to judge on ethical matters when the judges themselves are dragged before an invisible bar by other "judges"? And who has the right to punish the violators of moral laws? When we try to answer these questions, we enter into the darkness of unending disputes. One thing is certain: there is no one institution upon which we all rely in order to create and administer ethical laws. There is, for example, no supreme court of ethics before which one can defend an ethical judgment *finally*.

If you were to put those two eloquent men, André Trocmé and Adolf Hitler, in a comfortable room for any comfortable length of time to argue about the appropriate ethical judgment to lay upon the actions of the Chambonnais between 1940 and 1944, neither would persuade the other that his judgment was the final one; nor would they ever both acknowledge any court to decide conclusively between their conflicting judgments.

The Chambonnais judged the laws of Vichy in their own ethical courts, their consciences, and found them wrong. Vichy, for instance, did its best to enforce "surrender on demand," a decree that refugees must be given up when the appropriate state officials demanded it. In those airy, invisible courts of the human spirit, their consciences, the Chambonnais judged and condemned that decree, and they did all they could to violate it.

Still, it is not true that the public law and ethics are always on a collision course with each other. In many times and places they have worked together. For instance, in 1598 Henry IV of France issued the Edict of Nantes, which granted the Huguenots civil and religious rights far beyond any they had ever had in France. That edict created an era of comparative peace and cooperation not only between Catholics and Protestants but also between the consciences of Huguenots and the law of the land. The law and ethics are not natural enemies, one pursuing public order and the other committed to destroying it. When nations have been healthy, usually it has been because the two were in harmony.

Ethics is not only private; it can be communal. There are ethical communities, and one of them existed in the concentration camp near Limoges when the three Chambonnais leaders were there, despite the political and religious differences among the inmates. Such a community existed in the dining room of the presbytery when André Trocmé was arrested and his parishioners gave him some of their most precious possessions while Silvani sat there, cast down and weeping over untasted food. Such a community existed in the wintry, crooked street along the north side of the presbytery when the people of Le Chambon united their individual voices and minds to sing "A Mighty Fortress Is Our God" while the arresting officers took Trocmé away, with Magda Trocmé riding behind them. Ethical communities are as real as legal institutions. To call ethics personal is not to call it anarchic. It, like the law, can bring people together.

In fact, the more carefully one looks at particular areas in the law and particular areas in ethics, the more similarities one finds. For instance, both criminal law and life-and-death ethics seek to curb the destructive power of mankind. Both seek to restrain harmdoing; at least, both *claim* to be seeking this. In any case, where there is totally unrestrained destructive force, there is neither criminal law nor a life-and-death ethic. The "law of the jungle" is neither criminal law nor ethical law. Where might is right, there is no ethical or legal right.

In the process of restraining destructive power, both rely heavily upon facts. The law has rules of evidence, and in its own unorganized way, so does ethics. Anyone hearing Darcissac's or Trocmé's ethical judgments upon the people of Le Chambon would, as a matter of course, demand reliable testimony, facts, before assenting to those judgments. This book echoes their praise, but if I had no reliable evidence that the story you have read happened in Le Chambon, it would be sentimental nonsense for me to call the Chambonnais "good." And so, wherever possible, I have sought firsthand testimony (what the law sometimes calls "real" and sometimes calls "demonstrative" evidence). Without adequate evidence, both ethics and the law are empty.

2.

The law has fundamental principles—beliefs it assumes without proof, and frequently without comment. In American criminal law, it is presumed that an accused person is innocent until proved guilty beyond a reasonable doubt. The presumption at the foundation of life-and-death ethics is that all human life is precious. The Chambonnais showed their allegiance to this presumption by disturbing the center of their domestic lives and endangering their own existence to protect the lives of human beings, and in the process they never sought to do grievous harm to anyone.

The Nazis had other presumptions about the preciousness of human life. The official handbook of the Hitler Youth organization states those presumptions succinctly when it says: "The foundation of the National Socialist outlook on life is the perception of the unlikeness of men." Then it goes about developing the meaning of this assertion by pointing out that only "those of German blood" are precious, while the Jews are not. Indeed, it becomes clear in the course of reading the handbook (and any other major document of Nazism) that Jews are ultraenemies of "those of German blood." Military enemies have some value—for their valor, perhaps, or for the use one may make of them—but so different are Jews from those of German blood that they must be treated like disease germs or vermin, not simply conquered. The distance between this belief and the tenet that all human beings are precious is the ethical distance between the Nazis and the nonviolent Chambonnais.

The Latin root of the word *precious* is *pretium* (price). When we are talking about preciousness, we are talking about price. Like the phrase *in time,* the word *priceless* has two very different meanings: something priceless can be either very expensive or it can have no price attached to it at all, no equivalent thing that we will take in exchange for it. Diamonds are usually priceless in the high-price sense of the word, and children are priceless in the no-price sense of it. When a mother takes her child in her arms and says, "My precious child," she is not setting a high exchange value on the child. She is saying that she would not give up the life of that child for any price. When the Nazi guards threw living infants into furnaces in the death camps in order to save the price of a bullet or the price of enough Zyklon B gas to kill them, they were expressing a different belief about the preciousness of children's lives.

Under the moral leadership of André Trocmé and Édouard Theis, the people of Le Chambon would not give up a life for any price—for their own comfort, for their own safety, for patriotism, or for legality. For them, human life had no price; it had only dignity.

The American poet John Peale Bishop once wrote: "The most tragic thing about the war was not that it made so many dead men, but that it destroyed the tragedy of death." He was saying that in the course of the war, people lost their awareness of the pricelessness of life. They became accustomed to giving and taking human lives for a price. And he was saying that this was the worst thing about the war, that it destroyed the foundations of a life-and-death ethic. War substituted military heroism for dignity.

For those who believe in the absolute preciousness of life, there is no proof that the feelings and images involved in such a belief are correct or even plausible. They can only *show* their beliefs, the way a mother can show how precious her child is to her by rushing without hesitation into a fire to save it. Trocmé and the Chambonnais showed the value they put upon human life by four years of deeds.

From the time of Trocmé's first conversations with Burns Chalmers in Marseilles and Nimes, the saving of children was an important part of those deeds. We can begin to understand the idea of the pricelessness of human life if we explore the idea of the preciousness of a child's life.

When some of us look at children, especially when they are happy, we feel what the Jesuit poet Gerard Manley Hopkins was expressing in a poem called "Spring" when he asked: "What is all this juice and all this joy?" For us, children are the springtime, the creative burgeoning of human life. They are not only *in* that springtime; they *are* that springtime. Except when they are ill or hungry or in terror, their faces are fresh, free of the scars of old passions and enduring responsibilities. They seem to breathe their own excess of vitality into the things they touch or try to touch. Their surprise (and they are themselves surprises) sometimes makes us older ones want to draw close to them, the way a freezing traveler wants to stop his purposeful voyage in order to draw close to a warm and lambent fire.

And when they are tortured, when they are deliberately broken and killed, it is spring that is being attacked. It is as if the living center of human life were being dirtied and then smashed. An eyewitness (Rudolf Reder, as reported in *Betrayal at the Vel d'Hiv*) saw a mother and daughter at the head of a line going into the gas chambers of the concentration camp in Belzec, and he heard the child say, "Mother, it's dark, it's so dark, and I was being so good."[7]

Insofar as one can realize what is happening and what is evil about that murder, one is realizing the pricelessness of a child's life. There are people who do not comprehend what it means to be evil under a plain life-and-death ethic. There are some cold people on this earth. But there were no such people in the parish of André Trocmé. We may seek out, if we wish, the psychological, political, and even the economic causes of the differences between the cold ones and the Chambonnais, but as far as a life-and-death ethic is concerned, there is an unbridgeable difference between those who can torture and destroy children and those who can only save them. Under such an ethic, no verbal bridges may be erected connecting these two kinds of people. Between them there is only a profound conflict.

Of course, adult lives were precious for the Chambonnais. We who read these words can move toward an understanding of this preciousness by imagining and remembering the ecstasies of love, of food, and of music, in listening to which Nietzsche tells us: "the passions enjoy themselves." But we can also begin to grasp life's value by noticing our quotidian lives, apart from all ecstasy. How much money would we take in ordinary times in exchange for giving up the breathing, the eating, the drinking we usually hardly notice? Without noticing it, our bodies and our minds are usually celebrating a conviction deeper than words can express, a conviction that life is incomparably more valuable to us than death. When we are awake and when we are asleep, the heart in each of us pushes to beat, the lungs push to be filled.

In the course of writing this book, I had a heart attack. Afterward, day by day I learned as if for the first time what living is. During the first days after the heart attack, I felt distant from those who worked upon me and from those who visited me in the intensive care unit. I was aware of their healthily beating hearts, but with envy and with loneliness. The day after the attack, Michelena, my eldest child, in her terror, bent over to kiss me, and accidentally I seized her wrist in such a way that I could feel her powerful pulse under my fingers. I felt far away from even Michelena.

As the days went on, I began to perceive the healthy people around me with more and more joy. Now when I walk down the street and see a child who is as lithe as a cat, or an adult who is strolling with the unself-conscious poise of health, I rejoice with them.

Many survivors of horrors far greater than my heart attack have learned to say *yes* to life. Micheline Maurel, a survivor of the concentration camp at Neubrandenberg, emerged from infinite humiliation and suffering with a powerful belief in the blessedness of human life, of her life and the lives of others. Here is how she expressed that belief in *An Ordinary Camp*:[8]

> Be happy, you who live in fine apartments, in ugly houses or in hovels. Be happy, you who have your loved ones, and you also who sit alone and dream and can weep. Be happy . . . you the sick who are being cared for, and you who care for them . . . be happy, oh, how happy.

To realize the preciousness of life is to realize its value to all of us.

Such a realization, such an imaginative perception of the connection between the preciousness of my life and the preciousness of other lives, is the vital center of life-and-death ethics. If we do not discern that connection, the "laws" of ethics are empty patterns of sounds and shapes, without meaning or force. The moral leadership of André Trocmé consisted in keeping this perception green in his own life and in the lives of the other Chambonnais.

3.

Both life-and-death ethics and criminal law have codes. They make demands upon us that are more specific than the demands made by their presumptions. There are moral "laws" just as naked of governmental authority and power as ethics itself, but they exist somewhere, somehow. As Archbishop Saliège put it in his pastoral letter to the diocese of Toulouse, "They can be violated. But no mortal sin can suppress them."

But since there is no supreme court of ethics, and no other plain public authority to decide whether in fact those laws are being obeyed or broken, one of the most vexing questions in the history of ethics is: How do you know whether you are in fact obeying or breaking the ethical laws? How do you decide whether you are being ethical or unethical, good or evil?

The classic answer to this question is: Look into your own soul and see whether the laws are controlling your passions. Plato, Aristotle, the Stoics, Immanuel Kant, and the other makers of the classic conception of ethics in Western thought have told us this. In effect they have said that since ethics is not created and applied by an outward, public government, it must be thought of as a matter involving an *inward government.* When they are uncontrolled, our passions, our yearnings for sex and domination, for example, are like a rabble in a disorderly nation. They go wild, like an ungoverned people. By upbringing, by the word of God, by clearheaded thinking, by some means or other, each of us learns certain moral rules that help us to control our passions, to keep them in check, the way a well-governed people is kept in order. Ethics is inwardly experienced self-control. When the moral law within you rules your passions, you are good. When your inward government is in chaos, in anarchy, you are bad.

In short, ethics is only a matter of *character.* Aristotle, one of the founders of the classic conception of ethics in the West, tells us at the beginning of the second book of his *Ethics* that the word *ethics* comes from the Greek word for an individual's character. Virtue, he goes on, is a quality of the soul. The man of virtue, the man of character, habitually chooses the golden mean—never too much anger or fear or desire, and never too little. He moderates his passions and so gives order to his soul. And the crown, the reward—in fact, more accurately speaking, the very being, the very stuff—of that order is *his happiness.* You find out whether a person is good by finding out whether he or she is happy, whether he or she is enjoying a moderated, orderly soul. In fact, one's own happiness, Aristotle says in the first book of his *Ethics,* is "the best thing in the world," and is in fact "the end of all that man does."

Another great builder of the classic conception of ethics is Immanuel Kant. He opens his *Metaphysical Foundations of Morals* with perhaps the most memorable statement of an inward government ethic that has ever been written: "Nothing can possibly be conceived in the world, or even out of it, which can be called good without qualification, except a GOOD WILL." And he goes on to say that a good will is good "not because of what it performs or effects" but because it respects the moral law. This respect for the law means that the good person will not allow his own passions to overcome his reason and throw his inward government into anarchy. Kant is very careful to point out—and this care makes him one of the greatest figures in the history of the classic conception of ethics—that what we do to or for others is not central to ethics; the orderly condition of our own souls, our character, is what ethics seeks to achieve, and praises when it succeeds.

André Trocmé was a man of character, a violent man conquered by God, a passionate man whose respect for the Christian law of love controlled his powerful passions. But when I discussed his early life, I did not talk mainly about his self-control or about his own happiness at achieving that self-control. I talked about his actions in the world, about the actions of others around him and about how he and others interacted with each other. In that chapter and throughout this book, I have described his effects upon Célisse, Édouard Theis, Magda Grilli Trocmé, the Chambonnais, and the refugees who came up to Le Chambon; and I have described the effects of their actions and personalities upon him.

He was a good man according to the classic conception of good and evil, but he was more. Essential to his goodness, central to his decency, was what he did with and for other people, and what he did against them. In part he was good because he resisted the people who were doing harm and because he helped save the lives of those they were seeking to harm, the refugees. He was good because he diminished evil *in the world.* The evil he diminished was harmdoing, and the evil he diminished was suffering. His character was important, but it was not all, and it was not the only cause of the goodness he had and the goodness he did. Kindler, World War I, the comparatively lenient policy of the Nazis toward southern France at the beginning of the Occupation, the Huguenot tradition in the commune—these and many other elements were central to his goodness. These elements did not simply give rise to his goodness; they were at the center of that goodness as it happened in Le Chambon in the first four years of the 1940s.

When Darcissac and Trocmé praised the Chambonnais, they were praising not only what was happening within the souls of the Chambonnais; they were praising what happened in the presbytery, in the temple, in the boardinghouses, in the funded houses, and on the farms. The classic conception of good and evil as inward conditions of the mind or the soul is not totally wrong. It points up one of the important forces in ethical action: respect for the demands of ethics despite fear, despite indifference, and despite all the other passions that tend to debilitate and destroy action. In Le Chambon, at least, there was more to being good than this deep inward respect for the demands of ethics.

What there was is expressed succinctly on the certificate that accompanied André Trocmé's Medal of Righteousness when it was awarded to him posthumously by the state of Israel. That certificate described him as a man "who, at the peril of his life, saved Jews during the epoch of extermination." His righteousness was not dissimilar to the righteousness of those people he inspired and guided during the war years in Le Chambon. Ethical action is not isolated from history, as the classic conception has led many to believe. The study of ethics must not be afflicted with ecological impoverishment. It must not be a way of trying, by a use of abstract, traditional terms, to cast a fitful light within the inward worlds of men's souls. It must illuminate the great, rich regions of plainly visible human history. It must concern itself with the *story* of what individuals do in the context of the story of their times.

Novelists explore such regions of human history in their efforts to make us understand the ethical worth of a character. Charles Dickens, for instance, in *A Tale of Two Cities* has to make us understand a revolution and two great countries in order for us to grasp the ethical worth of Sydney Carton, and Herman Melville in *Billy Budd* has to make us understand the same revolutionary period and the same two countries in order to show us the ethical worth of Captain Vere, John Claggart and Billy Budd. Narrative, plot, and character, especially when the characters involved in the action are surrounded and pervaded by a world intimately involved in their deeds and passions, can help us to understand "good" and "evil" in large, clear, and concrete terms. And narrative can show us the many gray areas between good and evil, as well as the many differences of opinion about what kind of person or action *is* good or evil.

Because all this is so, the story of André Trocmé and Le Chambon tells us a great deal about the ethical meaning of Darcissac's term *wonderful,* when he applied it to the Chambonnais during the German occupation of France. Rich regions of human history as revealed in narrative illuminate ethics as much as ethics illuminates those regions.

4.

If ethics deals with such regions, what are the ethical rules that one may use to judge the rich actions of Trocmé and the Chambonnais? Whatever else they may be, they must be rules having to do with killing, rules involving matters of life and death, not rules having to do with sexual, professional, or other kinds of behavior that would be inappropriate to our concerns.

There are such rules. They are invoked by Archbishop Saliège in his pastoral letter to the diocese of Toulouse, and they were followed by Trocmé and the Chambonnais during those four years. They fall into two categories: negative commandments and positive ones. In an ethic of life and death, there is an ethic of refusal and there is an ethic of positive action.

The commandments in Exodus 20 are mainly concerned with a negative ethic, and as far as matters of life and death are concerned, the key negative commandment there is: "Thou shalt not kill." Kindler had put it in his own terms in Saint-Quentin when he said to the youthful André Trocmé, "One must refuse to shoot . . . and anything you add to this comes from the Devil." In ethical literature, certain other negative rules often accompany this, rules like "Do not hate" and "Do not betray." Dante in his vision of Hell divides that funnel-shaped pit into three areas containing those who did not obey these laws. Toward the top of the funnel there are those "who yielded to wrath," those who could not contain their passion to kill; these are the ones who have committed Dante's Sins of the Leopard. In a deeper part of the Inferno, farther down the narrowing funnel, are those who have committed the Sins of the Lion, the violent, those who have laid hands upon their fellowmen. In the deepest part of Hell, where the sinners are the most heinous, there are those who have committed the Sins of the Wolf, malicious fraud, or betrayal. Here at the bottom of Hell stands Satan, frozen forever at the center of the earth, forever weeping. It was he who committed the original act of betrayal by using language to tempt and betray Eve in the Garden of Eden, so that she violated the earliest of God's negative commandments: "Of the tree of the knowledge of good and evil, thou shalt not eat."

The Chambonnais did not believe in Dante's vision, not only because they had no doctrine of Hell, but also because they did not think that fraud was more blameworthy than physical violence or the passion that drives people to inflict it. However, they shared with Dante and many others the belief that fatal passions, destructive actions, and deadly fraud are wrong. By their refusal to hate Jews or any other human beings, and by their refusal to divulge the names of refugees in the commune to those who might do them great harm, they obeyed the main commandments of the negative ethic of life and death that has somehow come into being in the course of man's life on this planet.

The second kind of moral law that has come into being and has exercised an idiosyncratic, personal restraint upon the destructive drives of human beings is the positive ethic. Like the negative commandments, it has taken various forms, but these forms all revolve around the injunction to help those who are in mortal danger. The positive laws say, in effect: "Do something to prevent betrayal, hatred, and murder." The prophet Isaiah puts them in this way: "seek justice, correct oppression; defend the fatherless, plead for the widow." And Jesus explains the commandment to love one's neighbor as oneself with the story in the tenth chapter of Luke of the Good Samaritan who helped a man left half-dead by thieves.

According to the positive ethical laws, personal hygiene, clean hands that do no harm, and harmless passions and language are not enough; the decent person must have *working* hands, he must be his brother's keeper. He must do what he can to prevent others from violating the negative laws of a life-and-death ethic.

It is this positive aspect of ethics that the cities of refuge in Deuteronomy exhibit. When Moses tells Israel in Deuteronomy 19 to set apart such places of refuge, he makes it plain that if the Jews do not prevent innocent blood from being shed in these places, "the guilt of bloodshed be upon you." For citizens of such places, it is not enough to be harmless; it is necessary also for them to keep others from doing harm to those who come within the gates.

Édouard Theis has taken great pains to show me that he and Trocmé were trying to prevent the Nazis and Vichy from violating the commandment against killing. They were trying to protect the victims, but they were also trying to stop human beings who were hell-bent on becoming victimizers, hell-bent on doing evil. Trocmé and Theis believed that if they failed to protect those in Le Chambon, they, the ministers, would share the guilt of the evil ones who actually perpetrated the harmdoing.

Self-control and help are the two main aspects of a life-and-death ethic. Refusing to use our own power to destroy our fellow human beings and actively preventing others from using that power were the main ethical guidelines of the Chambonnais.

5.

It may appear that what happened in Le Chambon during the war years was almost too complex to be understood. In a sense, this is true. Looking at the story with an analytical eye, there were many factors at work before and during those years, forces that made the rescue efforts of the Chambonnais succeed. But if you are interested in understanding what happened in Le Chambon in a way similar to the way the Chambonnais themselves looked at what they did, then their actions become rather easy to understand. They become as easy to understand as Magda Trocmé rushing in her frenetic way from the kitchen to the presbytery door, turning the doorknob, and opening the door for a refugee with "Naturally, come in, and come in."

In physics the analysis of forces is useful. For instance, one may break down the various forces at work upon a door and upon the frame in which it is hung in order to hang the door well. But analysis is not all there is. There is another aspect to the full reality of this movement of the well-hung, opening door. There is the *experience,* so ordinary perhaps as to be unnoticed, of simply opening and closing a door.

If we would understand the goodness that happened in Le Chambon, we must see how easy it was for them to refuse to give up their consciences, to refuse to participate in hatred, betrayal, and murder, and to help the desperate adults and the terrified children who knocked on their doors in Le Chambon. We must see this, and we must also see the many elements that came together to make these things happen. Goodness is the simplest thing in the world, and the most complex, like opening a door.

We fail to understand what happened in Le Chambon if we think that *for them* their actions were complex and difficult. John Stuart Mill in his essay "Utilitarianism" wrote that a benevolent person is someone who "comes, as though instinctively, to be conscious of himself as a being who *of course* pays regard to others. The good of others becomes to him a thing naturally and necessarily to be attended to, like any of the physical conditions of our existence." For certain people, helping the distressed is as natural and necessary as feeding themselves. The Trocmés, the Theises, and others in Le Chambon were such people.

I have walked out of the temple after services with a crowd of peasants and villagers around me, and I have seen them one after another put astonishingly large amounts of money into the donation box to our left. Their clothes hung on their work-hardened bodies, and with the same clumsy ease that they exhibited in walking with those wooden shoes (one walks gingerly and high in them), they put their precious money into a box not for their own church but for flood or earthquake victims on the other side of the planet somewhere. Once I asked Theis, who was leaving with me, who these victims were (he had just put a very large bill into the box). He looked at me in surprise and said, "Oh, I don't know. They're people."

This wordless simplicity was important to the moving spirit of Le Chambon, André Trocmé. In an unpublished essay written in 1934 called "The Opposite of Evil," he expressed his belief that in times of crisis, theories and predictions are a refuge for cowards. In that essay he wrote of the dangers involved in trying to predict the effects of your actions on your own life, your family's lives, the lives in your parish, and the lives of your countrymen. During the war years he did not spend precious time and energy investigating the effects of his actions on any political theory he might hold. He chose to do without intellectual systems and without fear-filled predictions. He decided simply to "help the unjustly persecuted innocents around me." He decided to obey God's imperious commandments against killing and betraying. With his sophisticated mind, he put his sophisticated mind aside and chose to be a Kindler, who would not kill and would not betray.

6.

When he first came to Le Chambon in 1934, he noticed something curious about the terrain: it was difficult to get up to the village, but once you were there, you had what he called "the strange impression of being on a low plain." The great mountains and volcanoes that surround it in the distance and the gently undulating land of the commune give one the feeling of being at sea level.

The Chambonnais think, when they think about it at all, that they are at the sea level of human decency. If you insist upon discussing the matter, they will tell you that they are not morally better than anyone else.

But when I contrasted ethics with the law, I noticed that ethical judgments are idiosyncratic, personal: there are many different ways of judging deeds and people according to ethical "laws." One way of judging the Chambonnais is their way—with a shrug of the shoulders and the question: "Well, where else could they go? I had to take them in."

And there is another way of judging them, a way hinted at by the praise Darcissac and Trocmé gave them after the war. From the point of view of the Jewish and other refugees who walked or rode up that steep incline to Le Chambon, the Chambonnais are higher ethically than many other people. They are higher than those who during the war years lived in lazy indifference, or fear, or hatred. From the refugees' point of view, the Chambonnais are among the moral nobility of the earth, and the refugees know this as surely as they know the joy of being alive and as surely as they know the joy of seeing their children live.

Part of their joy, at least for some of the Jews, lies in their restored or newly born respect for Christians, at whose hands so many Jews have suffered pain and humiliation for centuries. And part of that joy lies in the refugees' renewed or new respect for our species, which contains such people.

There is another way of judging them in the idiosyncratic but indestructible language of ethics. There were the tears I shed when I first read about the khaki-colored buses standing almost empty in the square of Le Chambon. Now that I have examined the feeling behind those tears in the light of the clearest and highest ethical standards I have been able to find, the feeling is stronger and more useful in my life than it was before I examined it. I have expressed, to some extent, the deep beliefs of which those tears were a symptom. Those beliefs are personal: they are not to be found by looking around at our public institutions; in the end, they are to be found only in the dreadful solitude of one's own passionate convictions and doubts. They are to be found in one's own swift reactions to people and their deeds. They are insubstantial, but they do their work in us; and when they fail, we know, when we do not manage to deceive ourselves, that they have failed to work in us.

I, who share Trocmé's and the Chambonnais' beliefs in the preciousness of human life, may never have the moral strength to be much like the Chambonnais or like Trocmé; but I know what I want to have the power to be. I know that I want to have a door in the depths of my being, a door that is not locked against the faces of all other human beings. I know that I want to be able to say, from those depths, "Naturally, come in, and come in."

Postscript

1.

One evening in spring, Magda Trocmé came to dinner in our house. She came two years after I first read about the empty police buses in Le Chambon. It was so early in spring when she came that our usually colorful little Connecticut farm was drab. But she did not notice it. Immediately after she stepped out of the car, she drove her body and her attention toward my family and me. Despite her age and her rather recent serious surgery, she moved toward the door of our house, where my son and my wife and I were waiting for her, as if we were at the end of a low-ceilinged tunnel that made her hunch down and push frenetically toward us. I was standing in front of my wife and my son, and she greeted me with a look I had seen often in her eyes, a look that said, Well, we meet again, you cloudy-minded, excitable philosopher. You are seriously involved in my life, and you are in good health—well, then, let us *do* something together now. We do not have much time. And all of this was confirmed for me by the feel of her restless body when I took her in my arms.

I was surprised when all she gave to my lively, big-handed Italian wife was a shy, blank smile. I had told her about my wife's heritage of poverty in the hills of the Abruzzi. I found myself thinking, This visit will be disappointing. My daughter, Michelena, the moral plumbline, the militant conscience of our family, was away at college, and Magda would not meet her tonight, or perhaps ever. And she was not in harmony with my wife, Doris. But when my tall, sixteen-year-old son, Louie, took her right hand in both of his big ones—for two years, Louie's life had been full of stories about her—her interest suddenly awakened, and she opened up her face and her feelings to him immediately. He was about as tall as Jean-Pierre had been at the end, and as intense, with those eyebrows that met over his nose.

Before we sat down to dinner, my wife and I found ourselves talking about gardening in Connecticut with Magda's sister and brother-in-law, who lived in Connecticut and had brought her to us. During our conversation I found myself wondering where Louie and Magda were. After a while, I excused myself. I had Magda in my house, and I had some puzzling questions to ask her about the near-arrest of Jacques and André Trocmé in the Lyons railroad station. I found the two of them sitting on the sofa upstairs, leaning toward each other with intense interest. My son, with his mother's long, mobile face, was telling her something that was obviously of great importance to her. He stopped talking when I came close, and she leaned back hard with a light sigh of annoyance at the interruption. A little remorseful, but embarked, I touched my son on his big hands as I like to do, and I asked him what they had been discussing.

He looked at her for permission to speak for them both, and he said (without noticing how easily he used this great old woman's first name), "Why, Magda asked me how your book on her husband is going."

"Well, Luigi, what did you tell her?"

He looked into Magda's eyes for a little while, and then he turned to me and said, "Well, I told her that the book wasn't only about her husband. I told her it was about the village, too."

Out of the corners of my eyes I glanced at Magda, and saw her eyes glazed with remembering, as they so rarely are, and I saw in those eyes love for her remembered husband, and a little pride. In our many hours of conversation, I had often encountered her love for him and her loneliness without him, but I had never encountered her pride in him. She had been too matter-of-fact to praise him. She was perhaps getting sentimental at last, I thought.

"Well, Luigi," I pursued, in this mildly awkward conversation with my candid son, "did you tell her anything else?"

"Well, like, I told her that I know a lot about the book because our family is so close and we're always talking about it, especially at the kitchen table when we talk a lot about what we've been thinking. Then I told her about a lecture you gave on her husband and the village to the philosophy department at Wesleyan, and how, like, one of your best students said that he was 'perplexed' by the story of Le Chambon! *Perplexed!* Daddy, I told her that over there at the university they want to *make* things complicated. They want to make *you* perplexed. They want to break things down into little bits of nothing—nothing, Daddy! But isn't it just a true story, something that happened, and don't you simply admire the people of the village, and"—here he smiled at Magda Trocmé, who was now listening to him intently—"and, of course, Mr. Trocmé?"

I said, "Yes," and turned to Magda.

She was very happy. She said, "Oh, Luigi is a mature young man. Yes, mature." And she touched his hand with her own heavy one.

2.

We are living in a time, perhaps like every other time, when there are many who, in the words of the prophet Amos, "turn judgment to wormwood." Many are not content to live with the simplicities of the prophet of the ethical plumbline, Amos, when he says in the fifth chapter of his Book: "Seek good, and not evil, that ye may live: and so the Lord, the God of Hosts, shall be with you." Such round, plain words either bore us or perplex us, but in either case they are the beginning of confusion, not of understanding for us. A story like that of Le Chambon and an ethic that sees killing as evil and helping as good are too much concerned with the superficies, the obvious surface of things. The surface upon which light generously shines is not for us. We are afraid to be "taken in," afraid to be credulous, and we are not afraid of the darkness of unbelief about important matters. We are victims of the guarded wit.

But perplexity is a luxury in which I cannot indulge. With the Chambonnais I share a concern for timing. They did not have the time to speculate about the possible ways of formulating or applying ethical standards. They had to act not in due time but in time. I have more time than they had, and I have the advantage of retrospection, as well as other advantages. But never mind; I do not have enough time to postpone my praise of these people while I endlessly pursue such perplexing hypotheses as "What if the Nazis had won the war? Wouldn't they be the good ones now?" and "What if Metzger's Tartar Legion had attacked Le Chambon? Would Trocmé and the others have lifted a finger to save their lives and the lives of the children by fighting?" For me, as for my family, there is the same *kind* of urgency as far as making ethical judgments about Le Chambon is concerned as there was for the Chambonnais when they were making their ethical judgments upon the laws of Vichy and the Nazis. They and I, and you, have to judge events in time, before life itself sputters and goes out in each of us.

Those of us who have the luxuries of peace and retrospection have the time to examine moral judgments made in a time of crisis and great danger, but we dare not ignore the facts and the judgments that are evident in order to perplex our minds with endless arguments about what might have been. We must find what we can believe, understand it, and try to act upon it when the occasion arrives. I believe that we do not have time to mystify it. On matters of ethics we must see, understand, and choose our standards, or our lives are dark, though we may be patiently awaiting the light.

3.

I started the research on Le Chambon in solitude. Now the dead and the surviving Trocmés, Theises, and Chambonnais are part of a community in which my family and I live. It is not a community with palpable laws and a seat in space. And it is not a homogeneous community. It has atheists in it, devout Christians in it, and Jews. It is a community of ethical belief. One of the things we share is admiration for André Trocmé and what he lived for, especially in those war years. Another thing we share is the belief that for the most part the Chambonnais did the right sort of things during the war, though some of the things they did were amateurish, were not cunning enough to save precious lives. I have heard a Chambonnaise criticize Daniel Trocmé and Roger Le Forestier with tears of loving anger in her eyes for not being *careful* enough in those terrible times. Still another thing we share is the hope that if another crisis should come, we would with whole hearts do the kind of thing they did. And we share the hope that even in these less dangerous times we can do the kinds of things they did.

Solitude, estrangement from our fellow human beings, is part of our lives, as it is part of the lives of all aware people in our time, but it is not the most important part of our lives. Our awareness of the preciousness of human life makes our own lives joyously precious to ourselves. In the privacy of my home, and elsewhere in this ethical community, we have chosen that awareness as true north, from which we can take the bearings of our actions and passions.

For me, that awareness is my awareness of God. I live with the same sentence in my mind that many of the victims of the concentration camps uttered as they walked to their deaths: *Shema Israel, Adonoi Elohenu, Adonoi Echod* (Hear, oh Israel, the Lord our God, the Lord is One). For me, the word *Israel* refers to all of us anarchic-hearted human beings, and the word *God* means the object of our undivided attention to the lucid mystery of being alive for others and for ourselves. When I need commentary on the Shema in order to understand its meaning in practical terms, I recall Rabbi Hillel's summary of his belief in the preciousness of life:

> If I am not for myself, who is for me?
> If I care only for myself, what am I?
> If not now, when?

Notes

1. J. Glatstein, I. Knox, and S. Margoshes, eds., *Anthology of Holocaust Literature.* New York: Atheneum, 1973, pp. 375–381.
2. André Gide, *The Immoralist.* Trans. Dorothy Bussy. New York: Vintage Books, 1961, p. 125.
3. Alexander Werth, *France 1940–1955.* Boston: Beacon Press, 1966, p. 169.
4. Françoise Renaudot, *Les Français et l'Occupation.* Paris: Robert Laffont, 1975, p. 94.
5. Claude Lévy and Paul Tillard, *Betrayal at the Vel d'Hiv.* Trans. Inea Bushnaq. New York: Hill and Wang, 1969, p. 50.
6. André Trocmé, "The Law Itself Was a Lie!," *Fellowship,* January 1955, p. 4.
7. Lévy and Tillard, *Betrayal at the Vel d'Hiv,* p. 50.
8. Micheline Maurel, *An Ordinary Camp.* Trans. Margaret S. Summers. New York: Simon and Schuster, 1958, p. 140.

Sources

This is not primarily a book about books. Almost all the material used in this book is first-personal. It consists of testimony given by the people who helped make the story of Le Chambon happen.

The two most important sources are the unpublished autobiographical notes of André Trocmé, which he worked on until shortly before his death in 1971, and the spoken words of Magda Trocmé. The autobiographical notes are in the hands of the two surviving Trocmé children, Nelly Trocmé Blackburn and Jacques P. Trocmé, and in the possession of Magda Trocmé. The tapes containing Magda Trocmé's account of the story of Le Chambon are in my possession and in the hands of Nelly Trocmé Blackburn. A copy of them is in the Swarthmore College Peace Collection.

Aside from these two sources, there were many whose testimony is central to the book. Nelly Trocmé Blackburn not only gave me much material but checked parts of the manuscript for accuracy. Pastor Édouard Theis gave me hours of important testimony. Professor Burns

Chalmers told me much that only he knows, and so did Richard Unsworth and Daniel Isaac. Of people living in Le Chambon at the present writing, the following gave detailed testimony: Roger Darcissac, Monsieur and Madame Ernest Chazot, Madame Barraud, Madame Eyraud, Madame and Mademoiselle Marion, and Miss Maber. The tapes of the interviews with these people are in my possession.

Though these testimonies are far and away the most important sources of the book's materials, certain published works were influential in its writing and can illuminate the reader's understanding of the content of the book:

Bernard Gert. *The Moral Rules* (New York: Harper & Row, 1973). This book makes very clear the ancient distinction between negative ethics ("Don't kill" and "Don't betray") and positive ethics ("Prevent killing" and "Prevent betrayal").

Immanuel Kant. *Foundations of the Metaphysics of Morals.* Trans. L. W. Beck (New York: Bobbs-Merrill, 1959). This is an accessible, authoritative version of a "vertical," or commandment-centered, ethic.

John Stuart Mill. *Utilitarianism* (New York: Bobbs-Merrill, 1957). This is the best brief summary of a "horizontal," or benevolence-centered, ethic.

Friedrich Nietzsche. *Beyond Good and Evil.* Trans. W. Kaufmann (New York: Vintage, 1966). This and other works by Nietzsche seek to demolish the accepted ethical theories, and they throw a revealing light upon the nobility of André Trocmé, a light that the traditional ethical theories fail to throw.

André Trocmé. *Jesus and the Nonviolent Revolution* (Scottdale, Pa.: Herald Press, 1973). Part III of this book states Trocmé's views on nonviolence with great clarity and depth.

Two of my writings can help the reader to understand the basic approach of this book:

The Scar of Montaigne (Middletown, Conn.: Wesleyan University Press, 1966). Here I explain the "personal" approach to understanding knowledge, good, and evil. Such an approach involves bringing philosophy closer to literature than to the exact sciences, closer to the passions, actions, and common sense of individual persons than to a dispassionate, technical science.

The Paradox of Cruelty (Middletown, Conn.: Wesleyan University Press, 1969). This is a personal essay on the ways people entrap and crush people.

Certain books about France in the first four years of the 1940s can be read for an in-depth understanding of what the Chambonnais did:

Emile C. Fabre, ed. *God's Underground.* Trans. William and Patricia Nottingham (Saint Louis, Mo.: Bethany, 1970). This is an authoritative study of the rescue efforts of the Cimade, which contains a remarkable chapter on the Flowery Hill of Le Chambon.

Henri Noguères. *Histoire de la Résistance en France* (Paris: Robert Laffont, 1967–1976). Especially Vol. II, published in 1969. This month-by-month study of the Occupation is both encyclopedic and microscopic.

Robert O. Paxton. *Vichy France* (New York: Norton, 1975). One of the clearest and most authoritative studies of the Occupation in print.

Alexander Werth. *France 1940–1955* (Boston: Beacon, 1966). Part I of this book is an eloquent and basically accurate analysis of the period.

Index